Yesterday's Golcondas:
notable British metal mines

A scene perhaps never again to be witnessed. This photograph, taken about 1897, shows the removal of a 40 ton engine bob or beam from its house at the Cathole Lead Mine, near Mold, Flintshire. The 85in engine here was necessarily large in order to pump these heavily watered workings, and was supplied by William's Perran Foundry, Cornwall, about 1869. After dismantling, the engine was moved to a mine at Gwernaffield, also near Mold, an event which highlights the regular removal and re-erection of such machines which, despite their awesome size, was a common occurrence in the deep metal mining industry of the eighteen hundreds. Many large pumping engines of Cornish origin were at work in North Wales during this period, notwithstanding the construction of long, deep, drainage tunnels. The illustration provides us with a good idea of the size of these machines, which held the limelight for pumping purposes in all deep mining fields during the last century. Many of their neglected houses remain, open to the sky, occasionally visited by the curious, but otherwise silent and deserted. (Clwyd Record Office)   SJ206627

Yesterday's Golcondas

notable British metal mines

R. H. Bird

'For as birds are born to fly freely through the air, so are fishes born to swim through the waters, while to other creatures Nature has given the earth that they might live in it, and particularly to man that he might cultivate it and draw out of its caverns metals and other mineral products.'

Georgius Agricola *De Re Metallica*, 1556

MOORLAND PUBLISHING COMPANY

To the memory of my son.

ISBN: 0 903485 37 0

© R. H. Bird 1977

COPYRIGHT NOTICE

D
622.3409'41
BIR

Printed in Great Britain by
Wood Mitchell & Co Ltd, Stoke-on-Trent

For the Publishers
Moorland Publishing Company
The Market Place, Hartington,
Buxton, Derbys, SK17 0AL

Introduction

'The Tin Islands', or 'the land of the tin mines', were expressions applied by the civilized world to the British Isles more than two thousand years ago. While strictly misleading, as tin is to be found only in the south-western extremity of this country, these descriptions serve well to illustrate the fact that from very early times Britain was known to possess considerable mineral wealth. It is believed that the Phoenicians traded with the inhabitants of Cornwall and Devon, exchanging fabrics and other goods for tin and in consequence, it is probable that the Cornish 'mines' are the oldest in the country. At this time these mines would not be those underground workings with which we normally associate the word, for true deep lode mining was unnecessary, there being an ample supply of alluvial tin to be had at surface which could be relatively easily obtained by 'streaming'.

One of the prime reasons for the later Roman invasion and occupation of Britain was in order to exploit the country's mineral riches and while it is true that, particularly in the third century AD the Romans had considerable influence over the Cornish tin trade, their main interest lay in the mining of lead with its frequently high silver content, and also, to a lesser extent, gold, which occurs in localities in Wales. Here again, the extraction of lead would take the form of opencuts on the backs of veins, but gold was sought underground, due the concentrated but patchy occurrences of the deposits, as at the Ogofau Mine near Pumpsaint.

After the Roman withdrawal from Britain metal mining stagnated somewhat and it was not until the late sixteenth century that the great upsurge in mining activity, which was to place Britain in the forefront of world metal production and trade, came about.

The technical advances brought about by the Industrial Revolution enabled the deeper working of metallic ores, which had previously been inaccessible owing to flooding. These advances, as applied to mining, included the introduction of steam pumping engines. Such engines, although at first of the rather simple Newcomen atmospheric-type, were to put Cornwall in a world-commanding position and although tin continued to be produced there, it was copper on which the industry of the South West grew and matured. The other British mining fields were chiefly concerned with the production of lead, but in places in North Wales, Staffordshire and the Lake District, copper was concurrently produced. These mining fields were of a shallower nature than those of Cornwall and Devon and therefore the advent and perfection of the steam pumping engine had rather less effect on these regions.

The lead deposits of Wales and the Pennines had been continuously worked from very early times but with the growth of world trade, activity increased here at the same time as that of Cornwall, and production of lead ore (galena) and later zinc, soared.

The world has an insatiable appetite for metals and in the eighteenth and nineteenth centuries the search for other foreign deposits continued. At first these deposits, although very rich, were difficult to work due to their rather inaccessible nature and it was therefore some time before home production was affected. Indeed, shipments of foreign copper ores were initially brought to Swansea for smelting, but it was only a matter of time before huge deposits of the red metal near Lake Superior and the seemingly limitless tin ores of the Straits became, along with other sources, too much for the deep and ancient mines of Britain to compete against.

The lead industry on the other hand, continued without appreciable decline until about the 1880s and here again, exhausted lodes and heavy pumping costs took its toll, although there were notable exceptions to this trend with very important mines remaining in production well into the present century as, for instance, the Mill Close Mine in Derbyshire and the Halkyn Mines of Flintshire.

Bearing in mind that the industry is as ancient as that of agriculture and therefore production statistics can be little more than approximate guesswork, the output of metal from this country has been estimated as follows: metallic tin, considerably over 2 million tons; in excess of 11 million tons of copper concentrates which, taking an average metal content of say 7·5%, results in more than 825,000 tons of metal; 5 million tons of pig lead; $1\frac{1}{4}$ million tons of zinc concentrates and something in the region of 60 million ounces of refined silver. The countless mines whose combined outputs were responsible for the above figures were often little worlds on their own, most being conducted on the 'cost book' system by closely-knit groups of adventurers. Some were highly efficient concerns such as the London Lead Company whose care of its employees was the exception rather than the rule in the eighteen hundreds, while others were run with no regard for either the workmen or state of the mine and were orientated solely towards quick and easy pickings. All however had a common objective — to strike it rich. Some were fortunate in this respect, others less so, but it was a universal hope among both men and masters that the next shot fired would reveal untold wealth and prosperity.

While the industry is by no means moribund, the eighteenth and nineteenth centuries were generally the palmy days of British mining and the often quite feverish 'diggings and doings' have left their unmistakable traces. The gaunt and silent engine houses, the cavernous wheelpits, the adits, shafts and leats remain in varying stages of decay, each a tombstone to a mine of wealth both in aspiration and realization — a Golconda of yesterday.

The adage 'a photograph is worth a thousand words' is, from personal experience, never more apt than when applied to the currently popular study of Industrial Archaeology. With this in mind, I have attempted to compile a volume of photographs of Britain's non-ferrous mining remains in the hope that the reader will derive much pleasure and interest in the study of this erstwhile important industry, not merely by turning the following pages, but also by visiting some of the sites depicted because this is the only way of becoming properly acquainted with the fascinating subject of old metal mines and mining.

Each mining field has certain characteristics which differentiates it from the others, these characteristics being dependent on such things as local topography, longevity of operations, capital investment and modes of working. For instance, the 'old men' of Derbyshire, lacking the money necessary for extensive working of the ore deposits, were forced to use more empirical methods and equipment in their shallow mining operations which, although effective, were small in comparison to large companies working in the South West and the North Pennines. These differences are highlighted for instance in the contrast between the massive engine houses and dressing floor complexes of the West Country and the tiny climbing shafts and buddles in the Peak. These engine houses invariably contained the Cornish type of beam engine and it should be noted that the size of these was always given as the internal diameter (bore) of the single steam cylinder.

This is a book based mainly on photographs and as such, must necessarily be brief where technical information is concerned, but the reader is directed to a short selected bibliography at the back of this volume in which written works of both a local and general nature are listed appertaining to all those localities featured. I have also added a sprinkling of historic photographs both for interest's sake and also in order to give the casual reader a better understanding of some of the relevant remains which he may come across from time to time. It is always helpful to be able to locate sites quickly and for this reason each illustration is accompanied by a six figure Ordnance Survey grid reference, although it should be borne in mind that many mining sites often cover a considerably larger area around the given point of reference. As far as underground photographs are concerned, references are given for the entrance to the workings and it goes without saying that exploration of these should be left severely alone unless one is properly equipped, accompanied by an experienced person and the dangers involved are adequately understood. The inclusion of a date when the pictures were taken is considered a useful addition, in view of the deterioration which can take place both naturally and due to intentional demolition.

For those new to the subject, perhaps this collection of pictures may provide a guide and indication of what remains throughout the metal mining regions of England and Wales. Conversely, for those already well acquainted with this intriguing branch of Industrial Archaeology, the scenes recorded here may be useful in introducing an area hitherto ignored.

Like many other subjects, metal mining has a language of its own, and although technical words have been avoided as much as possible the use of some of them is almost unavoidable. To help the reader who may not be familiar with these terms a short glossary is included at the rear of this book.

Acknowledgements

The following persons have helped me in a variety of ways while this work was in preparation. Mr C. Williams of the Clwyd Record Office; Mrs D. Mortler and Miss E. Greaves; Mr H. M. Parker; Mr D. E. Bick; Miss J. L. Ratcliffe of Sheffield City Libraries; Mr P. Jackson and Mr P. Lloyd Harvey of the Mid-Wales Mining Museum Ltd. My friend Russell Bayles has kindly allowed me to reproduce certain photographs which he took in North Wales prior to the disappearance of a number of interesting remains and has also freely given information on steam engineering both in Wales and Cornwall as well as contributing most of the historical illustrations. Mr G. M. Rose and other members of the Peak District Mines Historical Society have made the task of underground photography considerably easier by their assistance with multiple lighting techniques, while my wife has made her contribution by stoically tolerating a husband almost constantly immersed in the subject of metal mining industry. Dr J. A. Robey, of Moorland Publishing has made many helpful suggestions as well as giving me the opportunity to publish material that was conspicuous by its absence in my first work. To all these individuals I extend my cordial thanks and appreciation.

Cornwall

1   The bleak and weathered engine houses of Wheal Edward bordering the cliffs above Cargodna Cove near St Just are somewhat overshadowed in fame by the nearby magnificently placed ruins of Botallack Mine. They are, however, well known, if not by their position, but by the fact that the workings here are interconnected with those of Wheals Drea and Owles, the former being the cause of a terrible disaster when they were accidentally holed after having been abandoned, and subsequently flooded, some time previously. The inundation took place in 1893 drowning nineteen men and a boy whose bodies were never recovered, since when the mine has remained unworked. The site has two noticeable features, the first being the smallness of the engine houses which contained engines of little more than 30in cylinder diameter — surprising in view of the mines' proximity to the sea, (the local granite and greenstone strata is notably impervious to water percolation so that heavy pumping was not necessary). Secondly, the existence of two inclined shafts, the Cargodna and Wheal Edward inclines, the latter still easily seen some distance beyond the left-hand house. This incline is sunk seawards to meet the 60 fathom level below adit and has an inclination of about 1 in 3. In addition to tin and copper, a wide variety of minerals are known to have occurred here including bismuth, uranium, pitchblende and arsenic.

(1974)   SW362329

2 Submarine mines such as Levant, were hampered in the handling of ore underground because steam whims (winding engines), which usually drew from surface shafts, were impracticable. This meant that all underground winding had to be accomplished by compressed-air winches whose power was limited. The problem at Levant was acute, for its levels extended some $1\frac{1}{2}$ miles out under the sea, and when a shaft was put down from the 178 to the 338 fathom level (called New Submarine Shaft) a powerful winding engine was erected in 1898 on the 260 fathom level for hauling through this shaft. To feed this engine and also to increase the air pressure for the rock drills used in the mine, a new compressor was installed at surface. Built by Holman Brothers, and set to work in 1901, it was a four cylinder cross-triple-expansion machine with cylinders of 17in, 27in and two of 30in, the whole measuring 60ft in length, and as such, the largest of its type in the Duchy. The management came under criticism from various share holders who considered that too much capital (about £8,000) had been expended on the machine, and

its ostentatious house. Judging by the ornate stack that still stands, perhaps their criticism was not unfounded, but to our modern eyes this interesting relic of Victorian industrial architecture is a refreshing change. (1974)

SW369345

3 In an attempt to smelt copper ores in Cornwall instead of being dependent on the more distant Swansea concerns, a smelting works was established in the 1750s on an extension to the tidal estuary at Copperhouse, near Hayle. During its heyday this smelting concern dealt with 4–6,000 tons of ore per annum, chiefly derived from the Camborne and Gwennap mines. Most of the smelting mill buildings and retaining walls of the wharf were constructed from square blocks of slag or scoriae derived from the smelting process and although traces of the smelting works have long since disappeared, the slag 'bricks' of the wharf walls remain. The large quantities of coal required for copper smelting allied to the existence of a powerful 'ring' of South Wales smelters caused the works to cease operations here during the early years of the nineteenth century.
(1974)   SW568380

4   The engine houses of the New and Old Engine Shafts at Trewavas Mine are not as well known as those of Botallack, St Just, although the former are just as spectacularly positioned on the cliff edge as at the latter site. When engines were being erected in their houses it was usual to heave the beam or bob into the house from the rear of the building, but due to the position of the nearer house on the very cliff edge with steeply rising ground at the back, manoeuvring of this huge part of the engine into position must have proved exceptionally difficult. Consequently the back wall was provided with an entry slot of large size through which to lift the beam, a feature which gives the house a peculiar appearance. While this arrangement undoubtedly weakened the structure, its continuing presence in this exposed situation speaks volumes for the skill of its builders, who were frequently called upon to complete engine houses in a matter of months, even in these hazardous situations. The mine was most active between the years 1836 and 1846 at which time it produced about 17,500 tons of copper ore, some of which was mined beneath the sea bed. (1974)   SW600265

5 The pumping engine and whim houses at the celebrated Cook's Kitchen Mine in the centre of the highly metalliferous Redruth-Camborne mining field. This old, deep and once rich mine was, like others in the locality, originally worked for copper and later for tin in depth. It was equipped for a time with a waterwheel-powered man engine. The property was worked until the beginning of the 1900s, being latterly incorporated into the Tincroft Sett. To-day the surface remains give little indication of the former importance of this mine and are notable only for the oversize masonry collar which surrounds the upper section of the pumping engine chimney stack.

(1974)   SW664406

7 Robinson's Shaft at South Croft mine in 1907 during the installation of a twin 22in horizontal cylinder winding-engine and two Lancashire boilers to steam it. The large engine house contains an 80in pumping engine (now preserved) which achieved some fame in being the last of its kind to operate in Cornwall, finally becoming obsolete in 1955. The engine beam is visible at the 'indoor end' of a stroke, while the tall shears — used in conjunction with a capstan for raising and lowering pitwork during maintenance — can be made out behind the headframe. This latter was replaced by one of steel construction in 1926 and the shaft still remains in use by South Crofty Ltd. Note the large pump barrels lying on the ground on the extreme right. Palmer's Shaft, on the eastern part of the property, is visible to the left while engine houses and stacks of the Carn Brea and Tincroft mines dot the landscape in the distance.   SW664411

6 Visitors to Redruth seldom notice
a small but once important building
near the town centre in Alma Place.
This was the mining exchange
founded in 1863, where much of the
local business of buying and selling
shares in the county's metal mines
was conducted.
(1974)   SW699420

8 The unusual building in the foreground of this photograph stands beside the road near Carn Brea Station on the site of the old South Tincroft Mine. This structure, with open-arched end walls, contained a horizontal tandem-compound single stage air-compressor manufactured by Harvey & Co about 1890, when the use of rock drills in the mine had been established. The engine was steamed from two Cornish boilers placed to the right of the building, these being served by a chimney stack of local style. The whim house beyond contained a 26in engine in use from about 1863 to 1894. In addition to winding duties, this latter machine also powered Tincroft's man engine, the rods for this being actuated from a crank with a 12ft throw geared down 4 to 1 from the whim crankshaft. This gave a working speed of four strokes a minute in a shaft situated 200yd WSW of the railway station. An interesting occurrence took place about 1894 prior to the engine's demise, when a fire destroyed a section of the rods in the shaft near the 160 fathom level. It is also reported that the man engine here had a rod break due to dry rot! In 1896 this mine was amalgamated with the Carn Brea group and continued operations until 1921 when it was considered entirely worked out. SW669408 (1974)

8

9 The interchange of coal and ore to South Wales from the mines near the Northern Coast of Cornwall necessitated the building of harbours of refuge for vessels engaged in this trade. While Hayle and Portreath still have harbours remaining intact, that at St Agnes has been reduced to a pile of granite blocks due to the pounding it receives from the Atlantic breakers. Originally constructed early in the seventeenth century, its position flush against the cliffs created access problems, and loading and unloading was therefore achieved by means of horse-worked windlasses and ore chutes situated on the cliff edge with a projecting timber platform for clearance. After many re-buildings due to storm damage, the harbour remained in use until shortly after World War I. Today remains of ore storage 'hutches' are visible on the cliff top, while below ancient workings on an adjacent tin lode are represented by many old men's levels which honeycomb the rock. (1974) SW721517

10 Remains of the storage 'hutches' above the old harbour at St Agnes. Two important mines in the locality which used this refuge for the shipment of ore and delivery of coal were Wheal Kitty and Polberro, these originally being supplemented by smaller concerns such as Polbreen, Gooninnis, and Trevaunance Mine. In addition to these, the mines around Perranporth are known to have made use of this harbour since there were no facilities for safe anchorage in Perran Bay. The engine house and stacks on the skyline are those of the Wheal Kitty sett around which the ground is pockmarked with dumps, shafts, and dressing floor remains. (1974) SW721517

10

11

12

11 and 12   Shaft sinking, due to cramped and ill ventilated conditions, has always been a particularly unpleasant business. In the early days of Cornish mining, shafts were frequently sunk in a most haphazard fashion, their downward course being governed by the dip or underlie of a lode. This made sinking very difficult and later, when pumping and hauling through such shafts, the extra work necessary in pitwork maintenance allied to the loss of ore stuff from ascending kibbles, became an increasingly expensive item in the working costs of a mine. The average size of most Cornish engine shafts in the 1860s was 8ft × 6ft, this size increasing in later years with the move towards multi-compartment shafts. Circular shafts were rather uncommon but at Dolcoath when the vertical New Roskear Shaft was commenced in the 1920s to prove the ground under the rich copper zone of North and South Roskear Mines, this was to take a circular form, 16ft in diameter. These two illustrations show the commencement of sinking and later, work near completion at a depth of 333 fathoms (almost 2,000ft). Note the waterproof clothing worn by the miners — protection against the constant deluge of falling water. The shaft was finally completed in April 1926 having taken 17 months to sink.
(1924) (1926)   SW654412

13   Compared with the beautifully stone-lined circular engine shafts to be found in other mining areas (particularly where sunk in shale and similar loose ground) those of the mines in the South West are nearly always rectangular and supported in weak ground by prodigious quantities of timber. It is also fair to say that the lack of good quality supporting masonry in other underground applications is conspicuous by its absence here, although concrete arching was erected in Dolcoath after major runs of ground had occurred. For this reason, numerous shafts in Cornwall are now in a very dangerous condition at surface, while below ground this fact is also readily appreciated. The photograph, taken underground in Polberro Mine, St Agnes, illustrates what can occur to shaft timbering over the years and in this instance, in a large stope, it would appear that support depends on the surrounding air alone!
(1974)   SW722516

14 Some fifty fathoms below surface in Polberro Mine, a tin mine re-opened this century through Turnavore Shaft which was deepened from 107 to 181 fathoms in an attempt to prove the West Kitty and Wheal Kitty flat lodes which were thought to run through the mine. A little work was done on the West Kitty lode but despite diamond drilling, the Wheal Kitty lode was never found. Polberro Mine is well known for the discovery and extraction in the 1750s of fabulously rich cassiterite deposits. Huge blocks of pure tin ore from 600 to 1,200lb in weight were raised, which without any dressing were transported direct to the smelters in Truro, and it was necessary to use carts for this operation as the usual pack horses were incapable of the task. The extensive timber work in the stope bottoms is in a dangerous condition and support hundreds of tons of waste rock over the level.
(1974)   SW722516

15 In order to prevent water from the drainage adits sinking into lower workings and therefore necessitating re-pumping, frequent use was made of wooden launders to bridge areas of loose ground. Sometimes these launders were supported on chains slung across gaping stopes and shafts and must have been a spectacular sight in times past. Present day exploration often reveals such ingenuity of drainage techniques, such as here in a side working in the Polberro Mine, St Agnes. The laundering of drainage water in loose ground and in surface leats is by no means unique to Cornwall and has at times been employed throughout most mining areas.
(1974)   SW722516

16 To support the sides and hanging walls of the cavernous underground stopes in the Cornish mines large quantities of timber were required. The timber took the form of massive square baulks of imported pine and were known as stulls. These forests of timber were often the cause of prolonged and crippling underground fires, such as those which occurred at St Ives Consols and Dolcoath Mine in 1844 and 1903 respectively. The most impressive use of stulls was often deep underground and are thus drowned and inaccessible, but an idea of the arrangement of such work can be gained in places near the surface as in the No 2 Level at the Cligga Head Mine near Perranporth.
(1974)   SW737536

17

17 A glimpse of underground life in the deep levels of a Cornish mine about 1890. Here, at the head of an underground shaft or winze, a small Holman air winch draws kibbles of ore from below the 248 fathom level in Tresavean Mine. The hard hats, worn by most underground workers, were useful locations for lumps of clay into which was stuck a candle, their only means of illumination. Many mines in the area obtained candle clay from a deposit near St Agnes, this material being noted for its extreme glutinosity. A bunch of candles or 'tallow dips' is visible hanging from a peg, in the top right of the photograph.
(c1890)   SW720394

18 The massively proportioned engine house and stack at the 170 fathom Batter's Shaft of the West Chiverton Lead Mine, is possibly one of the finest examples of Cornish mining architecture that remain. The wide ground-level arch shown in the illustration is unusual by being at the side of the building instead of the rear. It provided access for the 80in diameter cylinder and valve gear of the pumping engine during erection, which was built by Harvey & Co in 1868. This machine was described as one of the most perfect engines ever made in Cornwall and worked in this house from 1869 to 1882. The ore was rich in silver with some $1\frac{1}{4}$ million ounces of this noble metal having been refined from the galena, as well as a considerable tonnage of zinc which was recovered from the dumps at a later date.
(1974)   SW793509

19 The aptly named Red River which divides the towns of Camborne and Redruth has its headwaters near Newton Moor, a bowl-shaped depression south west of Carn Brea. The numerous large engine houses that stud the landscape here are an indication of the heavily watered nature of the terrain and the importance of former mining operations in the vicinity, which is crossed by the Great Flat Lode. This rich source of copper, and latterly tin ore, was worked by a number of mines, one of the more important groups being Grenville United. Looking to the south west from the elevated burrows surrounding Marriotts Shaft at South Wheal Frances, no less than thirteen engine houses are visible, all within about a mile from this point. In the foreground stand the whim and pumping engine houses on Pascoe's Shaft. As late as 1917 both houses were occupied, the latter with an 80in engine which, in that year was completely wrecked when the piston rod cap fractured, necessitating an urgent replacement of the entire engine except the bob. The replacement was 'foreign' to Cornwall as it was supplied by Worsley Mesnes Iron Works at Wigan, it being the last 'Cornish' engine ever to be built. Above

Pascoe's house on the skyline is Great Condurrow Mine, while hidden behind the whim are three houses belonging to West Frances and South Condurrow respectively. Discernible to the left in the picture are seven pumping, whim, and stamps engine houses situated mainly on Grenville United Mines. Still at work in 1919, this group boasted three large pumping engines (90in, 80in and 60in), but formidable increases in incoming water became too much for even these giant machines and mining ceased here in the following year.
(1974) Camera position   SW680395

20   The engine and whim houses on Fortescues Shaft at Grenville United Mines are an impressive size and well worth a visit. The 90in Harvey-built pumping engine (left) came from Tresavean Mine and started work here on 5 November 1892. After the mines' closure, she was sold and re-erected at New Cook's Kitchen Shaft, on property now owned and worked by South Crofty. The railway leads from the bank, over trestles and off to the stamps and dressing floors. The whim engine, drawing through the shaft is to the right.
(c1912)   SW668387

21   An interesting picture showing the sinking headgear and stack at Taylor's Shaft on East Pool Mine. Taylor's Shaft was commenced in 1922 and was urgently required to replace East Pool's main engine shaft which had suffered a complete collapse the previous year effectively preventing further work in the northern section of this mine where its future lay. The shaft was sunk to 283 fathoms and a large 90in engine was erected for pumping purposes. The sinking headgear was replaced by a much larger permanent one, some of the massive timbers for this being visible in the foreground, while the house for the 90in engine was erected the following year. The stack, which still stands alongside this house containing the preserved engine, remains a familiar sight to travellers on the A30 road at Pool, the white lettering on its crown denoting 'East Pool & Agar Ltd'. Note the masons' scaffolding for raising brickwork and the hole in the chimney side to allow entry into the upper section of the stack which is nearing completion.
(1922)   SW675416

22 A view of the twentieth century reworking of Great Condurrow Mine. The engine bob is hidden by the headframe. Note the massive balance bob on the right loaded with stones and used to assist the engine in raising the very heavy pitwork at each stroke.
(c1912)   SW662393

23 Visible for miles about, the well preserved engine house of Great Condurrow Mine dates from a reworking during 1907-13 when this house contained an 80in engine. This was the same engine which had originally worked at West Chiverton Lead Mine, from where it was moved to Gwennap United Mines until sold in 1905 to Great Condurrow. The granite-built house was originally to the right, but was moved to its present position on the other side of the 280 fathom shaft when it was discovered that the original location was sited upon weak ground. Although the Condurrow sett had originally been rich both in copper and tin, the twentieth-century reworking failed after only a small quantity of tin had been produced and at a time when the mine had just been unwatered. The four Cornish boilers of the engine were subsequently sold to East Pool and Agar for steaming Taylor's 90in engine, while the engine at Condurrow was broken up about 1915.
(1974)   SW662393

24 No work devoted to the history of national metal mining, however brief, would be complete without reference to the Great Consolidated Mines of Gwennap which, along with Devon Great Consols, were one of the richest and most productive sources of the red metal in the British Isles. Although worked during the eighteenth century they gained their notoriety during and after the time they were restarted in 1819 by John Taylor, whose London based co-adventurers placed considerable capital and faith in the unwatering of the area, a gamble which paid off handsomely. Several large pumping engines were employed for the initial re-draining of the mines (90in, 90in and 70in), these subsequently being added to as the workings went deeper and became more extensive. The massive copper lodes here were unusually hot, with air and water temperatures reaching 110° F in the aptly named Hot Lode, so that the miners had to work virtually naked and occasionally dowse themselves with cool water to enable work to continue. The large-scale ore production from Consols was the prime reason for the construction of the unique 4ft gauge Redruth & Chasewater railway, built to obviate the need for shipping copper via the monopolistic Portreath tramway and harbour, which was supported by Taylor's rival John Williams of Scorrier (nicknamed the 'King of Gwennap'), to whose chagrin the adjacent Consols were reworked. Due to Taylor's success, certain envious parties, in hand with the mineral lords, prevented a renewal of the lease which expired in 1839. As a result of this and prior to the expiry date, Taylor proceeded, perhaps with a little justified bitterness, to 'pick out the eyes' of Consols in a professional manner, thus leaving little for his successors. A new shaft was speedily and accurately sunk to facilitate the raising of ore previously blocked out and held underground in reserve, this ore tending to create a flood of copper at the monthly 'ticketings' (ore auctions). In 1861, Consols, along with United, Wheal Clifford, and other adjoining mines, were acquired by Clifford Amalgamated, a huge combine whose use of a vast array of steam engines for

pumping and winding has bestowed onto them the 'fame' of owning the greatest assemblage of such equipment ever to operate on a mining sett throughout the world. Water pumped by these engines — eight over 60in cylinder diameter — was derived from well over 80 miles of workings often 200 fathoms or more deep, and this containing high concentrations of corrosive acid, often quoted as sufficiently strong to rot a pair of boots off a man's feet within a day'. The Clifford Amalgamated Mines survived until 1870 when the lodes were found to be insufficiently rich to warrant the heavy cost of continued pumping. After the decision to abandon the mines had been made, all saleable materials were brought to the surface, and the fires in the boilers of the big pumping engines were drawn, permitting the water to begin its inexorable climb in the workings, and copper mining in Cornwall on an intensive scale slipped into history. Today the area is devastated with acres of deads and crumbling shafts, while most of the big engine

houses have been demolished, a fate shared by a large clock tower which stood near the count house at Consols. The surface remains therefore offer little to the historian and it is only by aerial photography that the true extent of this famous run of mines can be properly appreciated. The area of greatest despoliation, on United Downs, is shown in this picture looking NE towards Twelveheads. The conspicuous house (left of centre) was the last built on these mines and contained a stamps engine. The ruinous house beyond was constructed for Hocking's 85in pumping engine and this is one of the last houses left standing which contained the larger engines on the mines. Just visible in the top right-hand corner is Nangiles Mine where an 80in engine was installed and whose house perches above the Carnon Valley, through which ran the metals of the Redruth & Chasewater Railway. *(Cambridge University Collection : copyright reserved)* From SW743410

25   A general view of West Wheal Seton, situated to the north west of Tuckingmill between Redruth and Camborne. This shows the stamps and dressing floor of this notable working which, like so many mines in the locality was built upon copper and in later years turned to tin in depth. While a very successful copper mine, producing 125,770 tons of this ore between 1848 and 1890, the attempts at producing tin were bedevilled by inadequate dressing facilities and, due to the low prices of the latter ore in the 1880s, the adventurers were unwilling to re-equip the floors and the mine was sold up in 1891. The photograph portrays the somewhat decrepit state of the surface works before closure. Note the stamps engine with its secondary bob working a small lift of pumps to supply water to the stamps grates.
(c1885)   SW650415

26 A view through the rear wall arch of Wheal Jenkin engine house on Caradon Hill. The granite cylinder bed-stones through which the holding down bolts passed are visible on the floor, while in the distance can be seen the unusual engine house of South Phoenix Mine which has undergone conversion into a dwelling. Wheal Jenkin house bears a date tablet M.V.-1886-M.V. denoting Marke Valley Mines of which this was a part.
(1974)   SX265712

27 Because the Phoenix United mines were in operation at the beginning of the twentieth century, coupled with the fact that the area is in a remote part of Bodmin Moor, the surface remains are well preserved. This engine house is on the Prince of Wales Shaft and contained the last large pumping engine to be built in Cornwall for a Cornish mine. Viewing the remains here today one cannot help but agree with some contemporary critics who considered that the operation was conducted on too lavish a scale. Indeed, some £140,000 was expended on the sinking of this new shaft, but the operation was discontinued before effective exploration of the mine could be achieved, due mostly to the intervention of World War I and consequently the lack of further capital. The house with its square based stack and magnificently dressed cylinder bed stones for an 80in engine is the finest of its kind in the area. Old photographs exist showing the engine in situ and sinking operations in progress.
(1974)   SX26672.

28 The monastery-like ruins of the boiler house adjacent to Prince of Wales Shaft engine house. Grass now grows on the floor, although it is still possible to see depressions where the Cornish boilers once rested. One of these boilers was sold to the Basset Mines in 1915, although the engine was retained here in working order until the 1930s in possession of the Duchy of Cornwall.
(1974)   SX266722

29

30

29 A group of 'bal maidens', whose job it was to break down ore from the mine, are pictured here on the spalling floor at the South Frances section of the Basset Mines. The long handled hammers seen in the photograph were used for the reduction of large stones containing copper ore and then, when small enough, the ore was further crushed by flat-headed hammers or 'buckers' in the same manner as at the lead mines in the Pennines and elsewhere. A bal maiden's life was far from comfortable as she was frequently employed on the open floors in all weathers with very little protection from the icy winter elements. At the time this picture was taken, about 1895, she would probably receive from one shilling to one and sixpence a day.

SW679395 (approx)

30 Mechanical crushing or stamping was applied both to tin and copper ores, although generally it was the former which required reducing to a consistency of sand in order to free the finely disseminated cassiterite from the rest of the vein-stone. This operation was preparatory to the complex business of gravity separation or dressing, so that today, the ruins of a stamps engine house frequently marks the site of a dressing floor. Here we see a typical floor, belonging to South Condurrow Mine. On the right the 30in engine is operating ninety-six heads of stamps — forty-eight on each side of the massive flywheels — while the round buddles and other dressing equipment lie below and to the left. South Condurrow Mine (a part of Grenville United) was taken over in 1897 by the Camborne School of Mines for instructional purposes and thereafter renamed King Edward Mine.
(c1895)   SW664388 (approx)

31 Dressing floors became more sophisticated as improved and modern methods were evolved. The old floors had not the ability to recover all the small, but valuable, fines and these were washed away into neighbouring streams and rivers giving employment to hundreds of small concerns who further treated the tin-laden waters to great advantage. The latter-day mines realized that much profit was wasted in this way and, with the aid of large mills in which were installed modern complicated dressing machinery, much of this wastage was reduced. High speed frue vanners and Wilfley tables (shaking mechanical concentrators) were found to handle the 'slimes' and sands very efficiently and with certain modifications are still in use today. The picture shows the frue vanners in Grenville United mill early this century. Such equipment needed little attention and thus fewer employees were required.
(c1910)   SW664387 (approx)

32 and 33　New Consols Mine,
formerly Wheal Martha, is sited to
the north of Kit Hill near the village
of Luckett in the Tamar Valley. It
was an unusual mine in many
respects, not the least in raising ores
which were so complex as to
virtually defy the normal gravity
methods of concentration, these ores
containing a matrix of tin, copper,
arsenic, wolfram, silver and galena.
The name New Consols was applied
to the mine in the late 1860s when it
was taken over with a view to
utilising chemical treatment of the
ores in addition to the normal
methods. At the same time, considerable surface plant was added
including an 80in pumping engine
on Phillip's Shaft; a 50in pumping
engine on Broadgate Shaft; a 50ft
diameter water wheel; a 24in steam
whim; two crushers of 28in and 12in
plus a 36in steam stamps installation.
Despite this and further expensive
surface works, the mine failed to pay
its way and the venture was
terminated in 1877. Most of this
equipment remained, however, as no
buyer was forthcoming and the
neglected engines stood in their
houses in a derelict condition for no
less than 61 years until 1938 when
they were regrettably broken up.
Had they existed today, their
preservation would undoubtedly
have been assured as two of them
were probably the last examples of
their kind in the world. The
illustration here shows the stamps
and engine which powered them, a
secondary beam driven by the sweep
rod working a short lift of pumps to
supply water to the stamps grates.
Opposite is the 80in pumping
engine, the beam of which rests on
an extended and strengthened bob
wall owing to the house originally
being constructed for a smaller 50in
engine. Note the top of the pumps
beside the main rod. (*Geoffrey
Ordish*)
(1929 and 1936)　　　SX387736

34　The north-east portal of the
Tavistock Canal Tunnel which
burrows beneath Morwelldown to
connect with the Tamar Valley. The
canal, brainchild of the then young
John Taylor, was constructed for the
cheap conveyance of ore, coal and
other heavy goods chiefly required
by various mines around South West
Dartmoor. It was hoped that the
bore would reveal 'blind' mineral
deposits under the hill and in this it
was successful, a lode being encountered immediately inside this portal.
Named Wheal Crebor, the copper

mine was worked through a number
of shafts, the most notable being
Incline Shaft sunk at an angle of 35°
down to the 54 fathom level. With
its yawning entrance, which is just
off the picture to the right, the
incline was the principal haulage way
of the mine, this being effected by a
waterwheel driven by the swiftly
flowing waters of the canal. The
dumps from the working, which has
been estimated to have produced
some 41,000 tons of copper ore, are
found a little to the east, near the
River Lumburn, into which the

33

34

...ine's drainage adit discharged.
...ntry to the mine is possible
...rough this, although it should be
...oted that the workings in the
...amar Valley area are occasionally
...angerous to enter due to the
...resence of deep ochre mud in the
...vels caused by the decomposition
...f pyrite, which is associated with
...any copper deposits of the district;
... feature very evident in the main
...rainage adit of nearby Devon Great
...onsols Mine. The tablet over the
...ch records the date of
...mmencement of the tunnel.
...972)   SX461723

Wales

35 The most important metal mine in Southern Wales is undoubtedly Nantymwyn, set in a remote valley some 6 miles north of Llandovery. Because of the nature of the terrain this mine was easily worked to a considerable depth through adits driven into the valley sides, the two most important ones being the Upper and Deep Boat Levels. Subsequently, sinking below these levels for continued ore production necessitated the erection of a steam pumping engine about 1880, the ruined house and stack of which remains beside the Angred Shaft high on the hillside. The mine is an old one and there is a tradition that it was operational in the reign of Charles I. Of its many lessees perhaps the most well known were Messrs Williams of Scorrier, Cornwall, who held the lease of this notable working from 1836 until 1900. As a guide to its importance the records show that ore sales from this mine amounted to some 80,000 tons and as these records are incomplete its richness can be said to have been phenomenal.
(1974)   SN787445

36   The beautifully finished portal of the Level Fawr at Pontrhydygroes lies just below the B4343 road to Tregaron. This adit was commenced in 1785 (as stated on the keystone) and drained the Logaulas, Penygist, Glogfach and Glogfawr mines to the south. As with similar soughs in Derbyshire the driving of the level took many years and its final length, over 2,000yd was not achieved until it had been continued by a number of different companies of mainly Cornish origin. Used for haulage as well as drainage, a dressing floor was constructed about its portal. The account house on the site continues in use as dwellings while a considerable volume of water still flows from the level, although the latter is now blocked by roof falls.
(1974)   SN739723

Plan of Level Fawr and the lodes it intersects. The adit is sometimes known as Probert's Level after the instigator of the scheme who was connected with many lead mining ventures in the area. Level Fawr was originally driven to unwater Logaulas Mine and was not carried through to the Glog group until the mid-1800s, by which time these southerly mines were virtually exhausted in their upper levels and it was subsequently found that the expected richness in depth did not hold out in reality.

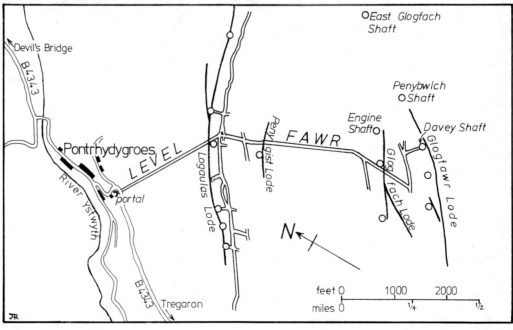

37   As at Esgair Hir Mine near the Nany-y-Moch Reservoir, the engine shaft at Penrhiw has a tunnel built into its collar, which housed an angle or balance bob for the pumps. Due to the removal of spoil in the immediate area, the shaft collar stands proud of the surface and could well be mistaken at first glance for a lime kiln. The shaft is sunk to about the 40 fathom level below adit, giving a total depth of 86 fathoms from surface. The mine worked the south-westerly extension of the powerful Castell Lode which here contains much iron pyrite waste mineral that still abound on the heaps. Along with other workings in the vicinity, Penrhiw has been amalgamated at various times under the titles, Aberystwyth Mines, Nanteos and Nanteos Consols, and has, with the others, produced more than 15,000 tons of galena and zinc blende together with a considerable amount of pyrite.
(1975)   SN737787

38

39   The collar of Vaughan's Old Shaft at Fron Goch. A pump rod, complete with iron strappings still remains in this 154 fathom deep shaft. The balance bob pits and lever wall of the engine house are visible, the latter having one wall in a completely collapsed condition. The house which contained the whim drawing through this shaft is slightly offset in angle while the waterwheel house is visible in the background. (1974)   SN723745

38   Foremost amongst buildings extant on lead mines in Cardiganshire are those to be found at Fron Goch. Centred about Vaughan's Old Shaft they consist of a pumping engine house, a whim house and a waterwheel building. The nearer house with square stack contained a 50in Cornish engine, and the middle building a 25in whim engine, both of which were installed in the 1860s so as to eliminate the mine's dependency on water power, previously used for pumping and winding. The cost of bringing in coal for these engines proved to be prohibitive, however, and the mine was latterly forced to return to water power for its operation. This was provided in part by a 56ft diameter wheel located in the structure on the extreme left. Originally opened in 1759 the mine was fully developed by John Taylor, whose skill and good management in metal mining ventures is well known and it was in his period of tenure that the present day remains here originated. After producing over half a million pounds worth of lead ore the mine changed hands and thereafter became an important source of zinc blende until final closure took place in 1910. (1974)   SN723745

40 The year 1899 saw the commencement of the final phase of reworking Fron Goch Mine which was undertaken by a firm of Belgian Lead Smelters, The Prayon Company of Liege. This company expended considerable capital on the erection of a dressing mill and power plant, the latter supplying high voltage alternating current generated by a Pelton wheel assisted by a steam engine, the whole contained in a large building beside the Cwm-newydion road, and whose ruins today resemble an abandoned monastery. A tunnel passes beneath the road, and through this was brought the steel pipes of the water main to feed the Pelton wheel; a dam was constructed in the hills above to provide the required hydrostatic pressure. Today the internal layout of the structure is difficult to visualise other than the position of the boiler room where the base of the chimney is in evidence in the centre of the ruins. Behind the building in the distance, the waste tips of Graiggoch Mine can be seen, above which the leat from Wemyss Mine is defined as a horizontal gash along the flank of the hill.
(1975)   SN706743

41 On the southern side of the valley of the Nant Cwm-newydion, the western end of the Fron Goch lode has been worked at the Graiggoch or Red Rock Mine. Originally opened by two adits, which have now run-in, the mine was developed by Messrs Taylor in 1840, who subsequently sank two shafts on the hillside to work the lode in depth. The principal pumping shaft was sunk directly onto the lode, and the pumps and hoisting machinery in this were powered by a 40ft diameter waterwheel. This wheel, together with a smaller one on the site, was fed with water conveyed from Wemyss Mine near the head of the valley, via a leat contouring the hillside, a feature still very evident today. The illustration shows the large wheel pit, beyond which can be discerned the grass covered tip from one of the original adit workings. (1975)   SN703740

42 Running diagonally up from the 40ft wheelpit to the engine shaft at Graiggoch are the remains of the dolly wheel stands which supported flat rods giving motion to the pumps in the shaft, sunk 60 fathoms below adit. Remains of such stands are unusual, and it is probably only due to the fact that the mine is in so isolated a position that they have survived. (1975)   SN703740

43 The large lead-zinc mine of Cwmystwyth has always been a favoured spot for mining concerns to try their hands, the Taylors, Aldersons and many others managing considerable profit and output over the years. The size and richness of the extracted orebody in both the Comet and Kingside Lodes can be appreciated by the cathedral-like stopes which remain accessible today. The considerable timber work supporting deads and the hanging wall illustrate well the complexities of underground carpentry. The last big enterprise here was engaged in the mining of zinc. This company, the Cwmystwyth Mining Company, was headed by Henry Gamman who sank much of his personal fortune into the undertaking. Unfortunately the quality of the ore was poor and taken as a whole the operation was not a success — the bonanza had passed.
(1976)   SN803746

44

44 The principal level, Bonsall's Level Fawr, gives access to extensive workings. After Gamman, Cwmystwyth was leased in 1909 by Messrs Brunner Mond who, despite large capital expenditure and development, fared little better than the previous company. Pleasant to relate, the plaque over the portal of Bonsall's Level which was removed some time ago has been reinstated and commemorates Gamman's spirited, but belated, efforts at this Welsh Golconda.
(1976)   SN803746

45 Due to the position of Esgair Hir Mine on a high ridge lacking a water supply for powering the pumps, a system of flat rods was constructed running up the steep hillside from a remote waterwheel sited in a valley to the south. This arrangement largely ameliorated the problem of drainage below the adit. The photograph shows the pit in which turned the pumping wheel. Large balance bob pits are sited at the far side of the area — so necessary since as well as giving motion to $\frac{1}{2}$ mile of flat rods serving Esgair Hir, this water-powered leviathan pumped another shaft at Esgairfraith to the north.
(1976)   SN743910

46 Almost the last large scale mining venture in Cardiganshire, the Bwlchglas Mine has left unattractive features at surface, the concrete hard-standings and machinery bolts contrasting with the now mellowed nineteenth-century mine buildings elsewhere. Bwlchglas was last operated by the Scottish Cardigan Mining Co, (with T. W. Wards of Sheffield having a major interest in the management) and was worked through two levels and a shaft. As with the Vieille Montagne Zinc Company's operations in the Northern Pennines, this company — which commenced operations in 1909—sank an underground shaft to reach the lower horizons of the ore deposits. The shaft, located at adit level at the end of an enormous stope, was equipped with two cages and a headframe which still survives, the power being obtained from a large diesel engine. Exhaust from this was carried out of the mine in ducting and heated the miners' 'dry'. The mine was lit by electricity generated by large gas engines at surface. On closure, many of the miners were transferred to the Penrhyngerwin Mine on the Dovey estuary, a venture also owned by the same concern. Although small companies are still working in the area, the closure of Penrhyngerwin just prior to World War II brought the centuries-old Cardinganshire lead mining industry to an end.
(1976)   SN710878

47 The installation of subterranean waterwheels for pumping and hoisting was a practice at many British metal mines, with large examples being noted at Cook's Kitchen, Wheal Friendship, the Ecton Copper Mines and Burtree Pasture Mine in Yorkshire, to name but a few; while overseas, the silver-lead mines of the Harz region had well over a score of these prime movers working below the surface. Although there may be others awaiting discovery, the only known example of an underground wheel in Britain is that at the insignificant little lead and copper mine at Ystrad Einion near Talybont. Here a 16ft diameter wheel was installed to drain a winze or underground shaft sunk below adit level to test the vein in depth. As well as pumping, it was connected through gearing to a drawing machine to hoist from the same winze and, despite a century's disuse, the whole remains in excellent condition and virtually intact. The installation of this and similar wheels was made possible by bringing component parts through the adit and assembling them within the cavern cut out to receive it. In order to keep dangerous amounts of water used to feed the wheel away from the shaft collar, where accidental flooding would be disastrous, wheels both at surface and underground were placed some distance away and worked through chains or flat rods — the former being applied to this example. A small balance bob (not visible) behind the wheel helped to equalize the load during the pumping cycle.
(1976)   SN707938

48 At the head of the underground shaft a length of bucket pump protrudes from the infilling water. The pumping chain is carried over an intermediate pulley stand between the shaft and the wheel. A small pulley sheave in the roof or on a timber headframe would have carried the hoisting rope.

49 The lower dressing floor of the Van Mine, Montgomeryshire is notable for its air of utter desolation. The huge banks of mill tailings, ruined walls and lack of vegetation are all that remain to remind the visitor that this was one of the most important lead mines of the British Isles. A lead deposit of extreme richness was discovered here in 1862 but it was not until the property was bought by the astute George Batters of London, whose West Chiverton Mine in Cornwall has already been mentioned, that the lode was properly developed. The vein was immense, being some 62ft wide in places, over a length of more than 2,000 fathoms. The main engine shaft was on a hillside above the trees where two stacks still stand. Three dressing floors dealt with the ore which were connected by an incline tramway, the latter reaching the lower floor through the tunnel seen in the illustration. Final working ended here in 1892 after more than 70,000 tons of lead and 25,000 tons of zinc blende had been raised.
(1974)   SN942876

49

50   Two stacks of the whim engine and 70in pumping engine remain on the top floor at Van Mine. The shaft through which thousands of tons of ore were hauled is now a deep rubbish-filled depression nearby. The lode here was of such a nature that large open stopes were undesirable because of the loose clay hanging-walls. Stopes had therefore to be immediately backfilled to consolidate the ground and as there were insufficient deads available for this work, special quarries were opened nearby to supply this material. (1975)   SN941879

51 Llyn Geirionydd near Llanwrst has been utilised for feed water to a turbine, which powered a lead dressing mill to the north. Latterly, and perhaps optimistically, known as the Klondyke Mill, this plant was fed with ore from the New Pandora Mine and other smaller workings around the lake by a tramway which terminated at the head of an aerial ropeway. This conveyed the ore some 200 feet down to the mill where it was crushed and dressed, a process which was aided, as at nearby Hafna Mine, by gravity feed. The large recessed structure seen in the photograph represents the lower end of the ropeway as is evidenced by a large iron pulley wheel inside. Opposite the mill is an imposing arched adit level bearing the date 1919. This was driven by the fraudulent Crafnant and Devon Mining Syndicate Ltd and was clearly intended as one of their major 'developments', but inside the portal two dividing tunnels end in poor ground after a short distance. (1974) SH765622

52 The levels driven from the Parc Mine near Llanwrst join with the workings of Hafna, Llanwrst, and Cyffty Mine to the south and are of considerable extent. The Principal Lode has been stoped away for many hundreds of feet giving the impression of the mountain having been vertically cleaved in half. The higher No 2 Level passes through this stoped area and provides an excellent view of the workings. Note the hanging wall and relatively puny stemples lodged across the stope of which only a small part can be included on a photograph. (1974) SH787593

53 Ore from this level was dropped down chutes (three of these being visible on the photograph to the right-hand side) to the No 3 haulage level below. Most Welsh mines were sited in areas of forestry and were thus able to make use of local timber supplies for underground support which was frequently in the form of cribwork, as in some of the mines in America. Cribbing comprises timber pillars built up in alternating layers similar to the construction of a log hut. While strong, it is also a wasteful method of roof support. Cribbing is visible in a minor form above the level, while the straggly and twiggy looking fungus so frequently found in this situation is also to be seen. (1974)   SH787593

54 The Llanwrst Mine still boasts an engine house and stack near the open engine shaft, as well as an angle-bob pit, thus indicating the use here of a short run of flat rods. The mine was of interest to share jobbers in the 1870s whose aim was the 'puffing' of mines whose fortunes were in name only. Such mining promoters were after quick profits from a gullible investing public, a practise which today would be considered illegal. To minimise this deception the surface buildings were constructed on a lavish scale to give the impression of great prosperity and it is probable that the remains here were a part of this ploy. The highest annual production was in 1878 when a mere 350 tons of ore were raised.
(1974)   SH779594

55 Underground at a point where the Parc Mine No 2 Level intersects a large engine shaft. The rising main of the pumps is clearly seen but the rods, activated by an engine at surface, have been removed.
(1974)   SH787593

56   Until 1966 there stood on the
site of Cyffty Mine, Llanwrst, a fine
whim engine house and boiler house,
together with an adjacent office
building and timber headframe over
the winding shaft. This compact
little group was needlessly demolished in that year, thus providing us
with yet another example of
thoughtless destruction of our
irreplaceable industrial heritage. The
house bore the date 1878 and
contained a steam whim of about
18in cylinder diameter. In addition
to steam power, a waterwheel of 35ft
diameter was used to drive crushing
and dressing machinery, its pit and
the remains of the supply reservoir
can still be made out amongst the
remaining ruins. Along with other
mines in the vicinity, Cyffty was
acquired by Great Challinor Mines
Ltd in 1914 and after a rather
uneventful working the lease was
surrendered in 1922, although the
workings were partially explored
from Parc Mine in the 1950s.
(Russell Bayles)
(1959)   SH775588

57

58

57 Only an engine house bob wall stands beside the 1,000ft deep Grosvenor Shaft of the Maeshafn Mine, near Mold. The oval lined collar of the shaft is still visible above the infilling rubbish. This once-fine house, situated on the eastern section of the Maeshafn Vein, contained an 85in Cornish pumping engine erected in 1865. The Maeshafn Mine was first opened in 1755 and throughout the middle of the nineteenth century enjoyed a boom period of production, latterly in the capable hands of John Taylor & Sons. (1974) SJ204611

58 On the western end of the Maeshafn Vein, set in sylvan surroundings, are two huge water-wheel pits. The wheel occupying the nearer pit was used for pumping via flat rods, while the far wheel probably drove crushing machinery located in or near the ruined building on the right. The arch visible next to the river was the exit way or tail race aperture. In addition to mechanical pumping, the workings were drained by an adit which discharges into the River Alyn nearby. (1974) SJ194614

59 An interesting feature of these pits are the two masonry 'chutes' in the eastern walling. These carried water below ground level to feed the wheel, thus making it of the breast driven type. The enormity of the wheel pits is best appreciated from within, as the photograph illustrates. (1974) SJ194614

60 North east of Llanarmon, two rich and powerful veins were being working during the eighteenth century: the Brynhaidd and Panty Gwlanod Veins. Sometime after 1850 a deep level was commenced in order to drain the workings here in depth. This, which became known as the Nant Adda Level, was driven in two stages, it being abandoned for some time until about 1890, when driving recommenced in conjunction with renewed interest in the area. The eventual outcome was disappointing and little in the way of production ensued. The neat portal is to be found near Cyfnant Bridge on the east bank of the River Alyn. (1974)   SJ187578

61 Lead mining in the Minera area is undoubtedly of great antiquity. As mining continued throughout the nineteenth century it was necessary to sink well below the water table to reach the most productive ore shoots. Because of water difficulties here, seven pumping engines were at work in 1817, raising between them 4,000 gallons a minute. Later, two drainage adits or 'day levels' were driven which enabled the mines to continue working profitably for many years. In recent times these levels have become a source of water supply for the district and one of them, the Deep Day Level, has been known to disgorge no less than $15\frac{1}{4}$ million gallons a day in sustained wet weather, thus giving an indication of the problems facing the miners. Furthermore, the standing water which now fills the workings below this Deep Level has been estimated as a phenomenal 135 million gallons. Minera Mines latterly produced large quantities of zinc blende, a mineral which often occurs below the productive horizons of galena and this mine became, in consequence, a most important source of that ore which found a ready market in the emerging galvanised sheet metal industry. Only one engine house remains on the site today and stands beside the 1,220ft deep Meadow Shaft. It contained a 40in pumping engine. (1973)   SJ275509

62　The mines of Flintshire were often exceedingly wet and in order to lay open rich ore deposits in depth, powerful steam pumping engines were employed. Two of the biggest were of 100in cylinder diameter, one at North Hendre Mine which was bought secondhand from Cornwall and started work about 1865, while the other was of more local origin and worked on Clive's Shaft at the Talargoch mine near Prestatyn. Built in 1862 by the Haigh Foundry, Wigan, this latter engine occupied the house illustrated, which remains complete with roof and weather boarding. The Talargoch Mine is very old but was most productive during the eighteenth and nineteenth centuries. Prior to the erection here of the 100in engine, Talargoch had employed a 50in hydraulic pumping-engine designed by John Darlington whose other similarly designed machine worked at the Alport Mines, Derbyshire. Closure in 1883 was primarily due to the heavy cost of coal (about £600 a month) used to fuel the fifteen steam pumping and winding engines. No less than nineteen boilers were in use to serve this array of steam power. Clive's engine was sold to a Wrexham colliery in 1884 for a mere £1,000. (1974)　　　SJ057804

63 The Halkyn Deep Level Tunnel was the first of two major co-operative drainage schemes undertaken in the Flintshire Mining field. Later extensions to this level discovered rich hidden ore deposits, the primary one of which (Great Halkyn Vein) was found 3,000ft beyond the Pant-y-gof or Deep Level Mine, which was the first mine reached by the adit from its portal. This blind deposit yielded well over 50,000 tons of galena together with much zinc. The tunnel was started in 1818 and was driven for nearly two miles before work was abandoned in 1822 due to the then poor demand for lead and also ventilation difficulties. Subsequently, work was re-started in 1878 by a company with the grand title, The Halkyn District Mines Drainage Company Ltd, whose use of rock drills enabled a driveage rate of up to 50 fathoms a month to be maintained. The level resulted in the unwatering of many hitherto flooded mines and was thereafter passing four million gallons of water a day and concurrently producing large quantities of ore from newly discovered veins. The adit eventually linked up with the Llyn-y-pandy mine near Mold in 1901. The deeper Milwr Sea Level Tunnel was later started in 1896 from Bagillt, by the Holywell-Halkyn Mining and Tunnel Co and superseded the first venture. As a result all the water issuing from the numerous workings is now taken by this later tunnel to be used by a large factory. Mining for lead in the area ceased as recently as 1958. However, interests in mining leases still exist in the area and the ground around the portal together with the Deep Level itself is private property and access to both is strictly barred. Below is given the information displayed on the plaque mounted above the portal arch.

This level was begun by the Right Honourable
Robert Earl Grosvenor
on the [?] day Nov'r. 1818.
Work recommenced and the level extended to Rhosesmor Mine
between the 19th day of February 1878 and the 7th day of July 1888.
Directors
His Grace the Duke of Westminster K.C. Chairman
Sir Thomas Gibbons Frost

John Scott Bankes	Col. Beaumont
H. R. Bowers	Robert Frost
Col. Heyworth	George Hughes

John Thompson
Engineers
John Taylor & Sons.
(Photo by P. Wild, courtesy of the Clwyd Record Office) SJ231710

The Great Holway Mine, also variously known as the Holywell or Boat Level was one of the most notable metal mines of the locality. This famous working was opened through an adit started in 1774 and constructed in the form of a canal tunnel to allow boats to be used for ore carriage, but was later drained and converted into a normal tramming level. A rich lode was cut in 1789 enabling the adventurers to glean over £160,000 in dividends until 1825. Steam pumping engines were widely employed as the workings sank deeper and the mine remained very profitable until the 1880s. The illustration shows a house that probably contained an 80in pumping engine during the time the mine was being worked by the Great Holway Lead Co which took over the lease in 1877. In common with many such structures, this building no longer exists, its site being cleared for modern housing developments and indeed, apart from the many miles of underground workings that remain, virtually all trace of this once important venture has vanished. Great Holway Mine has been the site of a number of steam engines including the 80in engine bought second-hand from New Pembroke Mine, Cornwall; two pumping and winding engines of 20in and 10in respectively; a 16in winding and crushing engine together with a 65in pumping engine working on New Engine Shaft. Latterly, Holway Mine was worked by the Vieille Montagne Co for its zinc content from 1901 to 1904. In 1919 pumps were installed in Roskell's Shaft to feed St Winifred's Well which had dried up as a result of the driving of the Milwr Tunnel from Bagillt.
(Russell Bayles)
(1965)   SJ179764

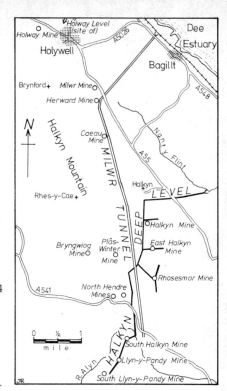

Plan of the drainage tunnel systems in the Holywell-Halkyn area. The true length of the Milwr Tunnel is not apparent on this plan, for it was afterwards extended to the Cathole Mine some ten miles from the portal.

65 Almost surrounded by trees and backed by a steeply rising bank stands an engine house on the North Hendre mine near Rhydymwyn. This house probably contained an 80in engine which, together with a second-hand 100in, had the task of draining these inordinately wet workings which were noted for the discovery of rich 'flats' and pockets of galena, one containing over 2,000 tons of this mineral; these deposits enabled the mine to continue operations until 1894. The engine house and adjacent red brick stack is one of the few remaining visible reminders of the once highly important lead mining industry of the area and this is a sad commentary on present day attitudes when one considers what used to be visible in the locality a mere decade or so ago. (1975)   SJ198678

Shropshire

66   Many obviously ancient workings are usually purported to be of Roman origin although often with little foundation for such claims. However, the relative proximity of the important Roman city of Wroxeter to the Shropshire orefield, allied to the discovery at the mines of Roman artefacts is perhaps proof enough that some workings here date from Hadrianic times. Indeed, in an openwork on the back of the later named Roman Vein, a pig of lead metal bearing Roman inscriptions was found at surface, while some 8 fathoms underground, nineteenth-century miners broke into workings containing tools, pottery and coins of that era. The opencut on Roman Vein does not exhibit typical early tooling marks, for the side walls are too shaley to retain them, but as this deposit outcrops on the valley slope and is therefore easily seen, it appears an obvious site for such ancient mining activity. (1975)   SO334998

67 The engine house on the Wood Engine Shaft at East Roman Gravels Mine is the sole survivor of four such structures which were built during the nineteenth century in the Hope Valley. Almost hidden from the main road below, this house together with the stumpy chimney stack some distance away, was erected by the Batholes Company about 1850. It housed a 30in pumping engine which was set to work by the West Tankerville Mining Co Ltd about 1870 — successors of the earlier concern — and pumped from 60 fathoms below adit in an attempt to unwater the mine prior to a reworking of the area. This venture proved a failure, and no worthwhile ore reserves were discovered on the property. The company struggled on until 1883 during which time it changed its name to East Roman Gravels Mining Co Ltd, continued sinking exploratory shafts together with the re-equipping of surface installations in the form of reservoir construction, new dressing plant, the erection of a compressor for powering rock drills and improvements in pumping layout.
(1975)   SJ335002

68 Hidden amidst undergrowth and totally out of sight from the main road some yards away, the portal of the Leigh Level remains as testimony to a once ambitious drainage project that failed. Designed to provide deep drainage for mines in the Hope Valley and the important Tankerville concern on the eastern outcrop 5 miles away, construction commenced in 1825. Due to legal disputes the project was abandoned ten years later only to be restarted in 1920 by Shropshire Mines Ltd who advanced the level to the 2 mile mark, at which point it was finally discontinued. Had this tunnel been completed there would have been little advantage to be gained as the mines it was intended to drain had already been sunk well below the horizon of the level. It was further hoped that the level would discover new ore deposits en route but the intervening ground proved to be entirely barren.
(1973)   SJ331035

69 A large engine house stands on the eastern end of the Grit Mines and dates from a period of working by John Taylor & Co in the 1860s. Although an astute mining company, the Taylor concern did not enjoy its usual run of success at this mine and abandoned it a few years later. No details of the engine are known, but it must have been quite large (possibly 30in or 40in) judging by its house proportions; it was employed for pumping, hoisting and crushing. The Wood drainage level reached the vicinity in the 1830s, but by that time the workings had sunk well below its horizon. Since abandonment by Taylor, the mine has remained disused and the 100 fathom New Engine Shaft in front of this building is blocked with over a century's accumulated rubbish.
(1975)   SO326980

70 Shropshire's most celebrated lead mine, Snailbeach, was responsible for over half the county's ore production. Supposedly worked by the Romans the mine's most active period began in 1782 and was to last for over a century. The present remains are largely derived from the second half of the nineteenth century when in 1858 George's or Engine Shaft had been sunk from Lordshill and a Cornish pumping engine installed. Subsequently the mine was provided with a new winding engine, compressors, a large reservoir, and improved ore dressing equipment. The narrow gauge Snailbeach Railway came into being in 1877 providing a link with Minsterley for carriage of ore and coal. The Snailbeach Vein ran through the mine for over 1,000yd and latterly realized 131,913 tons of lead ore, 4,392 tons of zinc, and was worked down to the 552yd level. The photograph from the dressing floors shows part of the crusher house, the chimney of the compressor house, and, on the skyline, the pumping engine chimney.

(1973)   SJ374022

72 Cliffdale Barytes Mill in the Hope Valley is completely obscured in summer by dense undergrowth. A determined search however, will be rewarded by a view of the Snailbeach Adit which emerges in this ruined building. Prior to the erection of the 61in Cornish engine on the mine, pumping was by a 36ft waterwheel erected here and driving a system of flat rods through the 1,200yd long adit. The adit joins with the 112yd level in the mine, the flat rod system working below this level via an angle bob. Later, in the days of steam pumping, the pumping rate was some 5,200 gallons an hour. Pumping ceased in 1910 and by 1913 the mine had filled to adit level. Explorations in the 1960s revealed that the adit is blocked by a fall some distance in, due to being driven mostly through shales and as water was being thrown some distance from the fall a high hydrostatic back pressure was assumed. (1975)   SJ364025

71 The two largest buildings remaining on the site of Snailbeach Mine are the engine and compressor houses. The latter was erected in 1881, this date being built in white brickwork into the chimney stack. Two Siemens and Edwards compressors were used to supply power to rock drills in the mine, together with pneumatic winches hauling through many of the winzes in the workings. After closure about 1895, the compressors were sold intact to Gresford Colliery and the building that once housed this equipment now contains only the massive mounting blocks, while traces of severed steam pipes are visible where the adjacent boiler house stood. (1973)   SJ375022

Derbyshire

73 Dirtlow Rake, Castleton is a good example of open-cast mining where the vein has been removed *en bloc* at surface leaving an open cut which can be followed for some considerable distance across Bradwell Moor. Due to the nature of the vein, the galena occurred close to the limestone cheeks or vein walls and in places the pick marks of the 'old man' can be recognised. Close inspection in these areas reveals specks and patches of the mineral, while the pick marks give an impression of great age due to weathering, unlike those to be found underground on occasions, which appear fresh, as though made yesterday. In places such open cuts are extremely narrow and in common with similar workings underground it is difficult to conceive how grown men could possibly work in such restricted conditions.
(1975)   SK154821

74 A view of what is thought to be an underground buddling cavern in Moorfurlong Mine, Bradwell. Sometimes where space allowed, ore was dressed underground which saved the labour of winding large quantities of material to the surface, the waste from this process being conveniently backfilled into old workings. Moorfurlong Mine worked a pipe vein which, as distinct from a narrow rake vein, swelled out into large cavities and here, near the base of the climbing way, much crushed waste material is to be found, obviously discarded at the dressing stage.
(1972)   SK168812

75 At the northern extremity of the powerful Dirtlow Rake near Castleton stands the limestone engine house and chimney stack of Ashton's or Pindale Mine. Little is known of this venture which worked the rake at a point where it dipped below the covering shales. The workings were drained by a sough, which was commenced about 1743; this has its outlet in the banks of a nearby river but it has now collapsed and is inaccessible. The mine's greatest period of production was during the opening years of the nineteenth century.
(1972)   SK162825

76

77

78

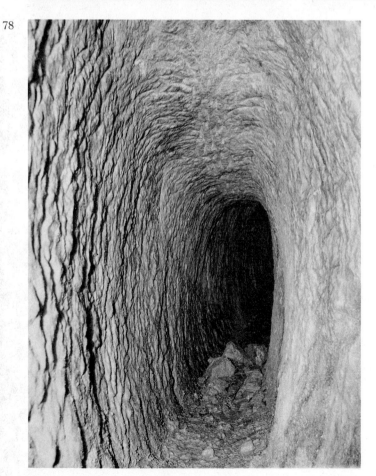

79 Great Masson Cavern is some 200ft long by 60ft in height and although not as large as some underground workings, its size is nonetheless, quite impressive. The tourist route runs from an adit driven in the Bacon Rake or Great Rake Vein and by way of a smaller cavern and passageway in which worked pockets bearing ancient pick marks are to be seen. Masson Cavern lies at the end of this system although this part of the mine is not currently open to public view.
(1975)   SK29358

5 and 77 Contained between the olid foot and hanging walls of xcavated rake veins, waygates or rifts in certain Peak lead mines did ot require to be fully arched. How-ver, the roof on which deads could e conveniently packed had to have ufficient strength to obviate run-ins, ore particularly over those levels hich acted as principal haulageways a mine. In the ancient and lebrated Odin Mine, Castleton, hich from early times proved very ch, dressed limestone stemples — gether with some drystone vaulting - have been employed over the ain Cartgate or drawing level, these

being used as an alternative to the more ubiquitous timbers with their inherent susceptibility to rotting. Constructed in the 1750s, Odin Cartgate has, in part, survived well due to this choice of roof support (the portal has however been lost as a result of nearby road improve-ments some years ago) and the old man's masonry work high in the roof remains firm and safe despite the level's floor having in places subsided some 40ft or more. Notice the hade or underlie of the extracted orebody — few veins or lodes are ever truly vertical.
(1976)   SK135835

78 A fine example of a hand-picked level through solid limestone exists within the Masson Hill Complex. These passages are known in modern cavers' terminology as coffin levels, due to their profile. This level is smaller than the nearby Founter-abbey Sough example in the Ball Eye Mine, and although it appears quite large due to the effect of wide angle photography, it is, in fact, only a mere three feet high by about eighteen inches wide. It is driven in part on a slight inclination and connects with a vertical hand-picked shaft leading to small upper workings. The typical sweeping pick marks in the roof and walls, together with the absence of shot holes provides us with a striking example of mining prior to the widespread introduction of explosives.
(1975)   SK291587

80   This illustration of High Loft
Mine depicts the enormity of the
workings which lie below Masson
Hill and shows a natural chamber
discovered by miners in 1779. Deads
or waste rock can be seen stacked in
any convenient place and many
places here have been entirely backfilled with this material so that the
true extent of the mine cannot now
be fully appreciated. The pool of
water is a typical feature of most
abandoned workings, and while this
one is very shallow, others found in
such locations may be flooded winzes
and stopes, many hundreds of feet in
depth and are a potential danger to
the unwary.

(1975)   SK291587

81 The cut-down structure of the Wakebridge Mine engine house near Crich. This contained a 60in Cornish pumping engine which was installed in 1857. Water was lifted by this to the Ridgeway Sough that enters the shaft 420ft from the collar. The engine ended its life at the Mill Close Mine near Darley Dale where it was kept as a standby until the 1920s. Wakebridge Mine has been sporadically worked since the deeper levels were abandoned towards the end of the nineteenth century, the last working for fluorspar taking place in the 1950s.
(1974)   SK339556

82 Near the Wakebridge Mine and adjacent to the road from Whatstandwell to Crich stands the timber headframe over the Rolley or Jingler Shaft. Although erected in the 1920s when the workings were producing fluorspar, this structure is unique to the area and has been preserved. The shaft here is sealed, but reached a depth well below the Ridgeway Sough which also linked up with this mine.
(1974)   SK340544

83 Thought to be the oldest free standing industrial chimney in Britain, that of the Stone Edge Smelt Mill remains a prominent landmark on the high gritstone moors between Ashover and Chesterfield. The London Lead Company were responsible for introducing the reverberatory furnace or cupola into Derbyshire in 1735, and it was near here at Kelstedge, that this type of furnace was first erected locally. This stack, built sometime before 1771, dominates the site while sections of condensing flues and other masonry work may be found around its base. A slag mill also occupied the site and the dam for supplying water to power the waterwheel-operated bellows for this lies to the east of the camera position. The records show that at the beginning of the nineteenth century, the mill was producing some 500 tons of pig lead annually, the ore coming chiefly from the Winster and Ashover mines. Stone Edge Mill ceased operation about 1850 and the fine square-section gritstone stack has recently been designated as an ancient monument.
(1974)   SK334670

84 Crushing and dressing machinery underground in the Mouldridge Mine. These isolated workings, in a south branch of Upper Gratton Dale, exhibit ramifying features typical of a pipe vein. The mine was last worked during World War II for fluorspar, although it was originally lead which attracted the 'old men' before that date. The equipment remaining from the last working was driven by a diesel engine, the exhaust from this escaping by way of an old man's shaft nearby. Since this photograph was made, the machinery has been removed by the Peak District Mines Historical Society, and can now be viewed at the Crich Tramway Museum where a replica 'mine' has been laid out for the benefit of visitors.
(1974)   SK19459

The Alport Lead mines represent a microcosm of a metal mining locality which had great problems of drainage and needed correspondingly ingenious solutions to enable much ore to be mined well below the natural water table of the district. In an area where coal for steam engines was prohibitive by its cost of transportation, these heavily watered mines made use of the copious supplies of surface water to power hydraulic pumping engines, together with an extensive system of drainage tunnels into which both the water from the engines and also that which was raised by them was delivered. Many early shallow drainage levels were constructed in the area, such as the Alport, Stoney Lee, and Shining Soughs, which permitted mining to lower horizons, but it was not until the arrival into the area of Hillcarr Sough together with its later branches, that deeper exploitation of the mines was facilitated. The level took 21 years (1766-87) to reach its objective at a cost of some £32,000. Of the many water pressure engines erected at these mines, the most impressive and famous was the Guy Engine, a model of which is to be seen in the Science Museum, South Kensington. With a cylinder diameter of 50in this engine worked in a shaft on Guy Vein, the latter now having entirely collapsed. Water was brought to this machine from the river Lathkill through an aqueduct level, after first being conducted over Alport Village in iron pipes. The portal of this tunnel was visible in the steeply rising ground to the south of the village, although debris from nearby poultry runs has since almost obscured it. As a guide to the work that this and the many other engines here were required to perform, it was estimated that between 2–6,000 gallons of water per minute entered the mines when they were operational. Today it is hard to visualise that beneath the green pastureland and limestone walls of the locality, a constant battle with water was waged for over two centuries.

85

85   The offices of the Alport Mining Company are today used as dwelling houses. On the left can be seen the Broadmeadow Shaft, denoted by a grassy hillock, the shaft beside it having run-in. Broadmeadow Shaft has housed two hydraulic pumping engines, the first being installed there about 1820, followed by a second machine in 1836, which, due to construction difficulties took nearly ten months to install.
(1974)   SK224644

86   The beautifully preserved Kirkmeadow Shaft is of singular interest. From the collar can be seen the fine ashlar stone ginging or lining disappearing into darkness, a feature which has prevented the shaft from collapsing under the pressure of the shale through which it is sunk. Also visible is an opening some 18ft down the south wall through which emerged pipes supplying a hydraulic engine located at the bottom. The shaft is sunk onto Thornhill Sough which drains Wheels Rake and is connected to Hillcarr Sough.
(1974)   SK23364

87 The gritstone walling at the collar of Kirkmeadow Shaft clearly shows the skill of the mason. The curved blocks are free of bonding so as to lock tightly together against lateral pressure of the ground. (1974)   SK233644

88 The outfall of Hillcarr Sough lies on private ground and is set back some distance from the River Derwent into which the water flows by way of a channel. Boats used in its construction were secured to the keystone when not in use, the ring for this purpose still being visible. Gritstone arching supports the level inside where poor ground was encountered, but this has collapsed some distance from the portal making the whole length now inaccessible. The atrocious conditions prevailing during construction were made worse by the prohibitive depths of the work below Stanton Moor preventing the sinking of air shafts, and it was not until the heading had passed beyond the high ground that such shafts became feasible. Mechanical methods of forced ventilation were used meanwhile, the air being fed to the forehead through lead pipes. On completion Hillcarr Sough was $4\frac{1}{2}$ miles in length and held the distinction of being the longest drainage level in Derbyshire and such was the success of the venture that construction costs are said to have been recovered within two years of its completion.
(1974)   SK258637

89 A major nineteenth-century smelting site was located at Alport. Owned by the influential partnership of Barker and Rose who had major holdings in the Alport Mines, this mill contained reverberatory furnaces and slag hearths, the fume from which was condensed in long horizontal flues prior to reaching the short chimney stack seen in the centre of the picture between trees. Although heavily outgrown and thus best viewed in winter time, this site is nonetheless one of the most complete of its kind in the county.
(1974)   SK223648

This plan of the flue layout at Alport Smelt Mill shows the intricate path indicated by the direction arrows taken by the fume prior to reaching the stack. A and B represent reverberatory furnaces, C and D represent slag hearths, while E is a calcining furnace. Condensers were incorporated in the system (shown as square and rectangular blocks) both between the slag hearth and at the base of the stack. These used water trickling onto stones and heather, as well as jets of steam to help condense the lead vapour. This water was conducted downhill from the top condenser via an intermediate one and ended in settling pits at Z. In addition, arched chambers 20ft long by 7ft wide and $7\frac{1}{2}$ft high were built within the long horizontal flue complex. Finally the vent to the 34ft chimney at O was supplied with a jet of water as a final 'scrubbing' operation. Reverberatory furnaces were used where large quantities of ore were available for smelting. The fire was separate from the charge to prevent the coal from coming into contact with the lead, the flames from the fire passing over a brick 'bridge' and were deflected or reverberated down upon the ore. A hopper was provided above the hearth for charging purposes; doors giving access to the hearth allowed the smelter to stir the melting charge and rake off the slag, the molten lead running through a tap hole. The slag hearth treated the resultant slags to extract the residue lead from them, and a high heat generated by water-powered bellows was required. Its operation, using coke as fuel, was similar to a blacksmith's hearth, the lead removed from the slags flowing out of a small tap hole into an iron pot. (J. Percy, *Metallurgy Of Lead*)

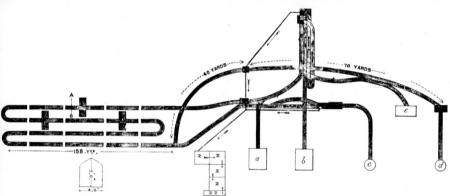

Alport Smelt Mill, Flue Layout

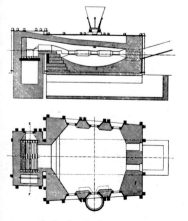

A. B. Cupola Furnace

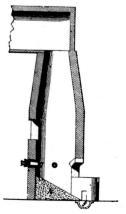

C. D. Slag Hearth

Isle of Man

90 and 91 The most important ores of this isolated island were lead and zinc with smaller quantities of copper being raised at a scattering of localities both within the main mining areas and also the south as, for instance, at Bradda Head. The mines were deep and extensive — Great Laxey, for example, having shafts sunk to some 300 fathoms — and in order to provide power for pumping, hoisting and dressing, water in the numerous streams draining off the high hills was harnessed to drive waterwheels and other hydraulic machinery. Steam power was used, but only at mines which were rich enough to erect and run the engines (the island had no indigenous coal supplies) and at locations where insufficient water was available throughout the summer to keep the wheels turning. Moreover, even at those concerns which were almost entirely dependent on water power — again Great Laxey may be instanced — long leats were necessary to supplement the streams and ensure a constant supply during the driest summers. Indeed, the island's mines were very similar to those in mid-Wales in this respect. It is therefore the features of water engineering which provide the most interesting and extensive remains at surface on most Manx metal mines. Great Laxey is, of course, justly famous for its massive Lady Isabella wheel and pump rod installation, but many visitors to the site are unaware of the great deal of surviving plant and buildings nestling in Glen Mooar beyond the Engine Shaft. The glen here is very narrow and since the main lode runs almost directly beneath it, the shafts and ancillary buildings were, of necessity, constructed within this narrow defile. While the stream was required for water power, it nonetheless proved an embarrassment at this point and consequently a bypass tunnel was driven to convey it around the machine houses and Engine Shaft.

The reformation of the Great Laxey Mining Co in 1854 saw the commencement of a period of expansion which, with small fluctuations, was to last until 1919 when the latter concern ceased operations. In the year of its reformation this company built the Lady Isabella waterwheel and concurrently refurbished the Engine Shaft. To hoist from this shaft, a rotative beam engine was originally installed, this hoisting kibbles of ore up to the Horse Level — later renamed the Engine Level after steam locomotives were introduced underground. The company replaced the beam winder with a water turbine system about 1860, but retained the former machinery as standby. The turbine was fed from a falling main connected to a cistern on the valley

side, the exhaust from this being conducted via launders to a lower cistern which in turn supplied the pumping wheel. Welch Shaft, situated a little higher up the glen was also wound by the turbine with ropes passing through the back wall of the machine house and via an arched culvert. The two photographs are taken from opposite sides of the glen, the right-hand picture shows the beam winding house with the turbine house just visible below; the view in the opposite direction shows the cistern and stone encased falling main to the lower house, while the northern portal of the Mooar Tunnel is just visible in the valley bottom to the right.
(1976)   SC432854

92 Part of the hoisting drums in the beam winder house. It may be noted that the gearing attached to the lower turbine equipment was provided with wooden teeth. Since turbine speeds are high, great damage would result should the kibble jam in the shaft while hoisting. If this occurred, damage would be localised to the destruction of the teeth which were quickly replaceable by the mine's carpenters.
(1976)   SC432855

93 As in Cornwall, the great depth attained by the Laxey Mine militated against the employment of the older and most experienced miners in the deepest levels, since they found it a great strain in climbing some 200 fathoms of ladders each day — a task which also wasted considerable time. In 1881 the company took the step of erecting a man engine in Welch Shaft, prompted, no doubt, by most of the mine's shafts being sunk on the underlie thus making cages difficult to operate. Man engines were a rarity in Britain outside Cornwall and Devon although one was installed at the Allihies Mines, Eire, in the 1860s. The engine, as may be expected, was of the water pressure type, having a 24in cylinder and giving a 12ft stroke in the shaft — sunk to the 295 level. The man rods were balanced by three balance bobs located in special chambers cut out of the shaft's sides; the engine made three or four strokes a minute lifting the miners from the 200 to adit level in 25 minutes as they stepped on and off platforms fixed to the reciprocating rods. Feed water was brought through an 8in main giving a 200ft head or some 18 tons pressure on the engine's piston. The main was bifurcated at the higher compressor house so as to feed both the man engine and compressor turbine from the same supply. This unusual arrangement proved very troublesome initially since the water hammer generated at every closing of the man engine's inlet valve badly damaged the turbine as well as rupturing an iron boiler connected into the system at the head of Welch Shaft. Thereafter a complicated series of weighted shock relief valves were incorporated to protect the higher turbine but, while partly satisfactory, these valves blew off at each stroke of the engine accompanied, we may be sure, by a hearty crash which reverberated through the supply pipe. On closure of the mines in 1919 by Great Laxey Ltd (the old 1854 concern changed to a limited company in 1903), the engine was abandoned in the shaft where it remains substantially intact — although slowly slipping from its seat of wedged boulders. Note the tilt of the cylinder due to the 15˚ underlie of the shaft. Water used by the engine was exhausted into a short crosscut connected with the turbine house from where it joined the latter's exhaust to be reused for the Lady Isabella waterwheel. (1976) SC432855

94 The deepest shaft on the mine was Dumbell's, sunk to 302 fathoms below adit — itself 41 fathoms below the surface at this point. Unlike the other shafts, Dumbell's was vertical and equipped with compressed-air pipes, telephone (a source of constant trouble) and hoisting facilities. The air compressors were sited in a building spanning the glen above the northern portal of the Mooar Tunnel so that the stream had to be conducted through the basement of the house in another tunnel. This is to the left in the picture while the right-hand arch gives access to the turbine which drove the compressors on the floor above. (1976) SC432856

95 A view of the compressor
turbine showing, left to right, the
rotor, transmission shaft to the
compressors, main shock relief valve
and below the figure, the bottom of
the falling main. Two further relief
valves are visible in the foreground.
(1976)   SC432856

96 The Foxdale and Great Laxey Mines were the two biggest metal producers on the island. The Foxdale group comprised a number of workings along the strike of the main E-W lodes together with those which developed the smaller deposits in the vicinity. The central mining area around Old Foxdale and Old Flappy Mines has little to offer the historian since most of the engine houses have been demolished and a major part of the site levelled. A clock tower still survives however, which was erected by the Isle of Man Mining Company as a memorial to its employees killed in World War I. To the west, on the ridge between Foxdale and Glen Rushen, stands the engine house of Cross's Mine, an obvious landmark for miles about. Foxdale Mines were heavily dependent on the power of steam, there being few usable streams on the property, particularly on the high ground. There was, moreover, a strong Cornish influence in these and other Manx mines as is exemplified by the many captains of Cornish origin employed there, so it was natural that a close liaison was maintained between the South West and Wales (another outpost of the Duchy) with regard to steam pumping engines. This attractive house and stack was erected for a 30in engine installed in 1840 and manufactured by John Taylor's Rhydymwyn Foundry, Mold. Prior to this, Cross's Mine had employed a 16-18in pumping and crushing engine bought in 1834 for £450 from Cornwall.
(1976)   SC263780

97 Over the hill from Glen Mooar at the head of the Cornaa Valley, lie the remains of Great North Laxey Mine which worked a lode running almost N-S below the valley bottom. Although not as productive as its southern neighbour it nonetheless was sporadically remunerative with a recorded output from 1856 to 1897 of 1,736 tons of lead concentrates. The most striking remains here are connected with the power source — again a large waterwheel — used for pumping the 174 fathom North Shaft some distance away. High masonry piers conducted water from the valley-side leat across to the overshot wheel, which was assisted in dry spells by a steam engine whose stack (with short ground flue) stands nearest the camera. This stack may have provided additional support for the overhead launders. Flat rods from the wheel were carried up the valley on masonry stands — seen to the left — and by way of a culvert to three further rising stands and so on to the angle bob at the shaft collar. It is probable that the siting of the wheel some way down the valley was in order to create sufficient fall since generally most wheels used for mining purposes on the island were of great diameter but of narrow breast. (1976)   SC428890

98 A general view of Great North Laxey Mine looking west. Two rising flat-rod stands are to be seen between the office ruins and the walled-round shaft collar. North and South Shafts are in line (the latter off the picture to the left) and between them is the site of a drawing machine with its wheel pit, this undoubtedly capable of winding from both shafts. The light area is a paved dressing floor containing the remains of a round buddle while in the foreground lies a wheel pit for the crusher.
(1976)   SC428890

99 Impressive surface remains are no guide to the richness of a mine. At Glen Roy considerable sums of money were expended without a halfpenny return to the luckless investors. The 50ft wheel here was fed by a long leat terminating in a line of high stone piers — visible from afar — the latter frequently having to be employed when using large diameter pitch-back or over-shot wheels supplied from negligible streams. This wheel has, of course, long vanished but its massive 'slot' or wheelcase remains in fine condition. By means of the usual flat rods, power was transmitted to the pumps in the very deep engine shaft located in the valley bottom; a smaller pit marks the position of perhaps a drawing machine or crusher wheel. Glen Roy Mine was opened in 1864 as a satellite of Great Laxey Mine and was later worked by a separate concern until 1882, by which time the shaft had reached a depth of 122 fathoms. Recorded output in the 1870s was a mere 9 tons 9 cwt of lead and 136 tons 9 cwt of zinc blende. The mine office still stands but the poor nature of the working is indicated by the insignificant dumps now clothed in undergrowth and pine trees.
(1976)   SC410835

100 Beckwith's Mine, worked by the Isle of Man Mining Co, lies on the extreme western end of the Foxdale group. Some 50,000 tons are recorded as having been raised between surface and the 70 fathom level in a vein averaging 10-15ft wide. This rather staggering figure, if correct, indicates the very bunchy nature of the vein at this mine, borne out later when bonanza followed borrasca with an eventual nipping out of the deposit in depth. The main shaft here is 185 fathoms deep with partial drainage through a day level. The workings proved to be very wet and much difficulty was experienced in keeping the water in check. In 1837 an engine from Coalbrookdale of unspecified size was erected which was followed in 1842 by a 50in engine bought second-hand from Gwernymynydd Mine, Flintshire, and in 1854 by a 70in supplied by Rhydymwyn Foundry, Mold. By the late 1850s however, the mine was in decline with an added disadvantage that great quantities of timber were required to keep open the levels, this timber needing constant replacement due to destruction by dry rot. The present-day remains include a fine washing-floor complex, crusher house, the engine house and stack beside the main shaft and a freestanding chimney which probably served the boilers of a whim. All the shafts — denoted by iron guard railings — have now run in.
(1976)   SC253778

101 A familiar sight to passengers on the Snaefell Mountain Railway are the white dumps that remain from a reworking of the old waste tips at Snaefell Mine in the 1950s. These, together with a lone chimney stack, a line of overgrown storage ponds and various concrete structures are a sad memento of this once well-known metal mine. Closer inspection reveals the water-filled main shaft (enclosed within a protective wall) sunk on the underlie to the 171 fathom level, and the collapse of which terminated underground activity here in the early 1900s. The mine was pumped by a typical large wheel, with steam in reserve, and mining starting in earnest about 1856. Between that date and 1900, over 4,500 tons of lead and more than 9,000 tons of zinc concentrates had been brought to grass (the greatest zinc producer on the island, and indeed Great Britain, was Laxey which produced some 203,647 tons!). Snaefell appears to have been an unlucky mine both in respect to its sudden demise and an accident in 1897 when twenty men suffocated to death as a result of a fire in timbering on the 130 fathom level.
(1976)   SC408875

02 The Isle of Man Mining Company was founded in 1828 by a group of Chester, Liverpool and North Wales businessmen, the managing partner being William Jones of the Buckley Lead Works, Mold. Included under the umbrella of this company the Townsends (sometimes known as Cornelly or Jones) Mine has the finest surface remains of the group. The site is dominated by a stack and engine house and here again we find evidence of Cornish links with the island. Into this tall house was installed a combined 50in/90in compound pumping engine 'similar to the engine now working at the Carn Brea (Cornwall) Mines, lately erected by the said James Sims'. The contract price was £1,900, the agreement with Sims being signed on 30 June 1841. It appears that the engine ceased work at this site about 1850 after which it was moved to Old Foxdale Mine. The mine lay dormant until 1874 when activity again commenced, but in common with many lodes in the area that at Townsends became increasingly poor in depth and closure finally came in 1884 with the shaft down to 140 fathoms. No output statistics are available since these were included with those of the Foxdale Mines, but it is stated that while considerable quantities of galena were raised, this did not cover the costs of working. (1976)   SC297795

103   Mining at Bradda Head is undoubtedly of great antiquity since the massive quartz lode exposed in the face of the headland would have attracted attention in early times. The mines of North and South Bradda have produced both lead and copper and have been in the hands of various entrepreneurs, none of whom seemed to have enjoyed startling results. South Bradda, sited at the base of the cliffs rivals — and indeed surely surpasses — Cornwall for coastal mine buildings. Here an engine was installed a few feet above high water mark to drain a 30 fathom shaft from which a deep exploratory drift was put out to test the lode — with disappointing results. The most productive workings were carried out by drifts driven into the cliff face, one of which is reputedly common to both South and North Mines. The employment of steam engines in this situation was not without problems, a major item being to secure pure boiler feed water. Mine water was tried but 'being brackish and intermixed with small spar it forms a crust in the bottom of the boiler . . . it requires a much greater quantity of coal to work the engine besides burning the boiler . . . .' The Bradda Mining Company were the last notable concern to work here and in the period 1881-3 raised 478 tons of copper ore. Two other companies who preceded them had raised 542 tons of lead and 339 tons of copper ore, an output, one suspects, that hardly covered costs.
(1976)   SC187696

Mid-Pennines

104 The bricked up portal of the Duke's Level in Hebden Gill, Grassington Moor. Driven between 1796 and 1824 at a cost of £33,000 this ambitious project was designed to drain the extensive area worked by the Duke of Devonshire. In this it adequately succeeded and provided drainage over a large network of levels and shafts to a depth of 72 fathoms. Cornelius Flint, the man responsible for the driving of this system, was at one time the Duke of Devonshire's agent at the Ecton copper mine in Staffordshire. (1973)   SE025644

105 Continuing up beside Hebden Gill beyond the portal of the Duke's Level, the track passes the remains of a paved dressing floor of the Hebden Moor Mines. From this point the Bottle Level was driven to cut the Cockbur and Star Veins from which most of the production ensued. (1973)   SE026651

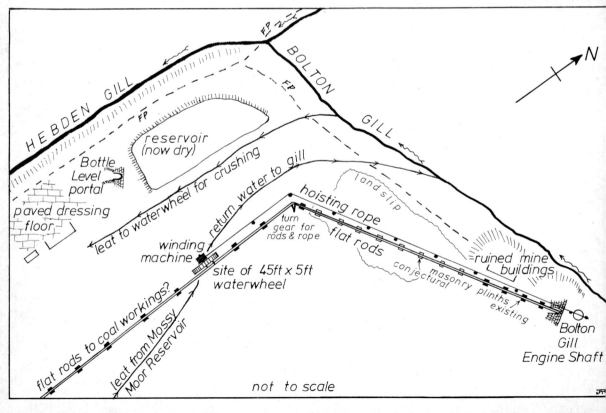

HEBDEN GILL

BOLTON GILL

FP

FP

FP

N

reservoir (now dry)

Bottle Level portal

paved dressing floor

leat to waterwheel for crushing

return water to gill

hoisting rope

land slip

turn gear for rods & rope

flat rods

winding machine

site of 45ft x 5ft waterwheel

conjectural masonry plinths existing

ruined mine buildings

flat rods to coal workings?

leat from Mossy Moor Reservoir

Bolton Gill Engine Shaft

not to scale

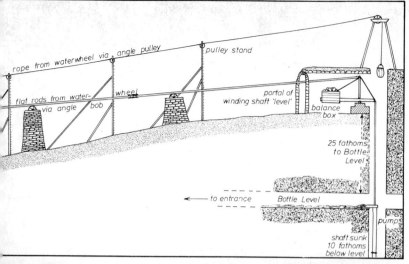

rope from waterwheel via angle pulley

pulley stand

flat rods from water-
via angle bob

wheel

portal of
winding shaft 'level'

balance
box

25 fathoms
to Bottle
Level

← to entrance   Bottle Level

pump

shaft sunk
10 fathoms
below level

Hebden Moor Mines. Sketch layout of the hoisting and pumping arrangements at Bolton Gill Engine Shaft. The waterwheel in Hebden Gill is thought to have also provided power for pumping at nearby coal levels and shafts which were sited about 500yd south and in alignment with the wheel. Due to this layout, the wheel-operated rods for Bolton Gill shaft were out of line for the latter and in order that both sites could obtain power from the same source, the engine shaft rods were turned through 59° via an angle bob sited 341yd from the collar. A similar arrangement was applied to the winding rope. Hoisting by waterwheel could not have been an important aspect of working the mine as most production came from the horizon of the Bottle Level and ore could be conveniently drawn through this directly to the dressing floor at its mouth. Water used to power the wheels was stored in substantial reservoirs in Bolton Gill and on Mossy Moor. The mine was worked from 1853 to 1890 by the Hebden Moor Mining Co who were also given the right to mine coal. Peak years for lead production were 1862 and 1863 when 425 and 352 tons were raised.

106   The Bottle Level meets an engine shaft sunk from the slopes of nearby Bolton Gill, and which was equipped with pumps and winding gear operated by a waterwheel in the valley below. Power from the wheel was transmitted via flat rods carried up Bolton Gill on massive masonry pillars and by way of an angle bob and balance weight at the shaft collar. This was contained in the large structure seen here, while the winding rope was carried over the usual headframe above. The Bolton Gill winding shaft has recently been restored by the Earby Mines Research Group.
(1976)   SE029654

107   Looking out from the top of the Engine Shaft at Bolton Gill showing the interior of the structure that housed the angle bob for pumping down the shaft.
(1976)   SE029654

108

108   At the dressing floors of the
Low Stoney Grooves Mine beside
the headwaters of Ashfoldside Beck
there exists a well preserved example
of a Cornish-type round buddle. The
fines or pulp, which constituted the
final stage in the dressing of metal
ores was fed into the centre of the
buddle via launders and this material
was stirred by rotating arms actuated
by a waterwheel. In this way the
heavier mineral (the metallic ore)
settled near the centre while the
lighter waste was distributed towards
the periphery, thus separating out
the valuable galena. Examples of
buddles in such an excellent state of
preservation as this are now unfortunately becoming a rarity. This one
is thought to date from the time
when the mine was being worked by
the Bewerly United Lead and
Barytes Mining Company in 1887.
(1969)   SE097667

109

109 and 110 Gillfield Level, Greenhow, was begun in the 1780s along with the Cockhill Level a little further west, and fifty feet higher. Both were the approach to extensive ore shoots which lie beneath Greenhow Moor. The levels are interesting in that they each contained in their inner sections, an underground boiler house, steam pumps and winding equipment, these being installed in an attempt to prove the ground below the level. Due to the chimney shaft collapsing in Gillfield, it was impossible to continue with the steam equipment underground and this deep exploration came to an end about 1840. Access to the inner reaches of the system is now no longer possible due to roof falls. Gillfield Level intersects, amongst others, the Waterhole and Sun Veins, the latter being represented by large stopes which rise to a height of 80ft above the adit. The timber props, which have been used in conjunction with side arching, are in good condition, although they support prodigious quantities of deads. The stopes were more recently the source of fluorspar when the mine was reopened in the 1930s, this mineral being milled in a small installation sited near the level mouth where the old smelt mill stands.
(1968)   SE116648

111 The arched stone canopies which housed two smelting hearths at Marrick Old Mill are probably, along with those in the higher New Mill, the best preserved example of their kind in North Yorkshire. Here the bellows were worked by a 22ft diameter waterwheel situated on the south side of the building, while the fume from the hearths was conducted through two parallel flues to a chimney situated at the New Mill building. When last visited by the writer, the grooved cast-iron work stones from the front of the hearths were still to be seen lying in the grass beneath each furnace canopy in the New Mill.
(1973)   SE079995

112

113 The higher reaches of Gunnerside Gill are well known for its mining landscape and there are many fascinating remains of the Swaledale lead industry. The gill cuts across the principal east-west ore shoots and has enabled the working of these to be conveniently carried out both by hushing and the driving of extensive horse level networks. On the western side of the gill a very large hush which is crossed halfway up by the track to Blakethwaite Smelt Mill, denotes the course of the Lownathwaite North and Middle Veins and is known as North Hush. Its great size is indicative of its having been worked for very many years before underground working replaced it.
(1975)   NY935014

112 A system of hydraulic mining that was long used in the Central and Northern Pennines was a process known as hushing, whereby periodic bursts of water were released from a dam constructed above suspected vein outcrops. In its downward rush the water removed soil and rock from the hillside, thus exposing the ore shoots. The area was ideally suited to this form of 'mining' as the ore deposits generally occurred high above the river valleys. Hushes which were successful in exposing the veins were frequently worked for many years and in consequence grew to enormous proportions. Some of the most spectacular hushes are to be seen around Swaledale and are found in great concentrations on the west side of the Arkengarthdale valley at the point of outcrop of numerous strongly mineralized veins. From north to south these comprise Dam Rigg Cross Hush, Stodart Hush, the Hungry Hushes, Band Hush, Dodgson Hush, Stemple Hush and the Turf Moor Hush. This photograph, looking almost due east, shows the bottom of Stodart Hush, perhaps the most spectacular of the group.
(1975)   NY986028

114 In 1881 the A.D. Mining Company installed a 12in hydraulic pumping engine and a 5½in winding engine in an underground chamber at the inner end of Sir Francis Level for the deep exploitation of the Friarfold Vein. The high rainfall of the area and steepness of the valleys was ideal for operating such machinery and, in fact, the nearby Blakethwaite Mines had employed a hydraulic engine some forty years previously in a similar situation. The Sir Francis engines were fed from the Sun Hush Dam on the moortop, the supply pipe 1,800ft long leading down the hillside to a 43 fathom shaft sunk to the engine chamber. This 13in diameter pipe has now disappeared except where it projects from the shaft, and this can be found near the stream about 40yd south of Priscilla Level. In the background are the tips from Bunton Level.
(1975)   NY938014

114

115

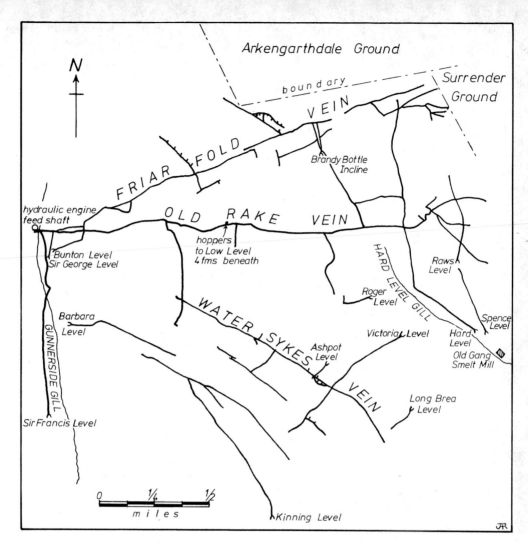

Arkengarthdale Ground

boundary

Surrender Ground

FRIAR FOLD VEIN

Brandy Bottle Incline

OLD RAKE VEIN

N

hydraulic engine feed shaft

Bunton Level
Sir George Level

hoppers
to Low Level
4 fms beneath

HARD LEVEL GILL

Raws Level

Roger Level

Spence Level

Barbara Level

WATER SYKES VEIN

Victoria Level

Hard Level

GUNNERSIDE GILL

Ashpot Level

Old Gang Smelt Mill

Long Brea Level

Sir Francis Level

0   ¼   ½
m i l e s

Kinning Level

115 A view from the west side of Gunnerside Gill looking towards Bunton Level. This level was designed to link up with the workings of the Old Gang Company which were driven from the parallel Hard Level Gill and which exploited, amongst others, the Friarfold and Old Rake complex of veins. John Davies, the agent responsible for the driving of Bunton Level appears to have had little ability in mining matters and perhaps due to faulty dialling (surveying) drove the level at too high an inclination (a feature quite evident when viewed from inside) so that it was necessary to sink a shaft 4 fathoms down to the Hard Level network to make the connection. Further, much of the ore the level was designed to tap lay beneath its horizon as a result. The level mouth lies behind the building, while the extensive ore storage bays (or bouse teems) to the right of this were seldom used after connection with the Old Gang workings had been made — all ore being tipped down into the lower system and trammed out of the Hard Level. The obvious trackway across the hillside led from the Sir Francis level mouth enabling the ore obtained from that mine (worked jointly by the Old Gang and A.D. Companies) to be brought up the gill and into Bunton Level, bound for the Old Gang Smelt Mill. The tip on the extreme right marks the position of the Woodward Level.
(1975)   Camera Position NY937014

A plan of the vein and horse level network at the Old Gang Mine, Swaledale. This plan — from an original dated 1888 — clearly shows the complexities of the workings which lie between Gunnerside and Hard Level Gills. The key to this ramifying system, the Hard Level, was commenced about 1785 when the mines were being worked by Lord Pomfret. Although seemingly on one plane, the level portals and crosscuts are actually at differing horizons.

Northern Pennines

116   Sited close beside the Alston to Middleton Road near the head of Teesdale is the Ashgillhead Mine whose surface buildings typify those mines located on the bleak and windswept Pennine Moors. The buildings here have been used for farm purposes and therefore remain in reasonable condition. The portal of Ashgillhead Horse Level is backed by a building universally known in the area as a 'shop' and would probably contain an office and smithy together with a room for material storage. In really remote situations 'lodging shops' would be built which contained sleeping accommodation for the miners who, due to the distance from their home, would be forced to reside at the mine for the week. Because the horse levels also drained the workings, the miners could expect to spend each day underground with wet feet and as lodging shops were both insanitary and over-crowded the prevailing conditions at such mines were incredibly poor. Ashgillhead Mine was owned by the London Lead Company who worked it until 1885, producing 4,622 tons of lead concentrates.
(1975)   NY808355

17 The Northern Pennine orefield was subject to considerable development in the eighteenth and nineteenth centuries by large concerns, notably the Blackett/Beaumont and London (Quaker) Lead Companies. The Beaumont family worked large tracts of ground in Weardale and Allendale, while the Quaker Company leased a greater part of Alston Moor from the Greenwich Hospital, as well as operating most of the Teesdale Mines. These companies had considerable skill and capital available, enabling them to undertake an intelligent programme of exploration and development which resulted in the discovery of rich and valuable ore deposits. The Nenthead area was the headquarters of the latter organisation and it was here that a large dressing floor (said to be the finest in the North) and a smelting mill were built. These were designed to deal with ores trammed out of an extensive horse level network nearby, which still remains as the greatest complex accessible in the British Isles. Of these levels, the most notable are Smallcleugh and Rampgill, which gave access to a system of metasomatic replacement deposits known as 'flats'.

Approaching the village of Nenthead from the direction of Alston, the visitor is confronted by huge dumps of mill tailings which, despite recent reclamation, still remain an impressive size. Continuing through the village (which was built to house employees of the company) and following the river, the portals of Rampgill and Capleclough Horse Levels are visible, above which is to be found the remains of Nenthead Smelt Mill. The mill, originally constructed in 1737 by a former lessee of the area contained four ore hearths and four refining furnaces. When the London Lead Company took over the mill in 1745, it was redesigned to the reverberatory principle — two of these furnaces and a slag hearth then being installed. The mill thereafter continued in use until the Quaker Co surrendered their Alston Moor leases in 1884. Since that time the building has been used by the Vieille Montagne Zinc Company of Belgium as a steam-driven compressor house during their term of zinc mining operations and latterly, as a dressing mill by small tributers reworking the old dumps of the locality.

(1975)   NY784433

118 A large paved area behind the Nenthead Smelt Mill is surrounded on two sides by what, at first sight, appear to be ore storage bays or bouse teems. These are in fact the remains of coal bunkers used in connection with the Vieille Montagne Zinc Company's reworking period, when the buildings housed Lancashire and Robey boilers to supply steam for compressors sited here which were in use until a hydraulic system, in the nearby Brewery Shaft, had been completed.
(1975)   NY784433

119 The flue from the smelt mill crossed the track via a bridge (since demolished) and wound its way up the moor to terminate in a chimney stack which still stands. In order to condense the fume more thoroughly, various experiments were conducted, and in 1843 Joseph Stagg, who was the London Lead Company's manager at Nenthead, patented a condenser which used a powerful pump to draw the smoke through tanks of water in order to precipitate the poisonous content of the fume. These pumps were driven by a large

waterwheel, the crank from which was connected to a bob working on the substantial masonry wall sited at the end of the wheel pit. The bob in turn drove the pumps. After initial troubles caused by acid corrosion, the condenser proved quite efficient and was thereafter installed at other mills, but construction costs often militated against its widespread adoption, the simpler Stokoe condensers using brushwood and water sprays being more frequently favoured.
(1975)   NY7854³

120 Rampgill Horse Level at a place now euphemistically referred to as 'Whisky Bottle Corner' — in reality a junction of diverging haulageways serving the complex workings of this notable lead mine. (1976)   Portal NY782435

121 and 122   Evidence of long-dead miners are occasionally found, such as here in the London Lead Company's old stopes above Rampgill Level. These workings are approached through an unusual arched rock stairway which rises from the main horse haulageway. (1976)   Portal NY78243

121

122

123 The portal of the Smallcleugh Horse Level appears unremarkable and belies its fame as the entrance to miles of levels, rises, stopes and flats. Beside the portal scattered stonework remains from the demolished level 'shop' and smithy.
(1975)   NY787428

124 and 125 One of the most remarkable features in the underground workings of Smallcleugh Mine is the beautifully made drystone arching. The greatest use is made of this in the flats, where the haulage levels pass through areas backfilled with deads, and because of the general dryness of the workings, the arching today appears as sound as when it was first constructed. The stone for this work was specially quarried on Flinty Fell and brought into the mine — the honey-coloured masonry contrasting sharply with the grey limestone of the workings — and was undoubtedly easier to split and dress than the deads in the mine. The illustration here shows stacked deads above the arch, one side of which springs from steel bars supported on metal pins. Opposite we see the vaulted and arched stonework — all held in place without the use of mortar — where the level widens out at a loop for the mine tubs to pass.
(1975)   Portal NY787428

126 and 127   A further two views of the arching in the Smallcleugh Mine. A report in 1822 on mines leased by the Greenwich Hospital stated: 'the walling of shafts, and the arching of the levels was another great improve- ment, many of the old excavations having fallen in, whereas now the mines are rendered accessible at any future period'. As far as this mine is concerned, there cannot be any doubt as to the wisdom of these words. Note the ore chute which projects out over the rails for ease of loading the waggons and, opposite, the skilled work where a passage branches.
(1975)   Portal NY787428

128   In order to reduce the depths of the winding shafts from the surface, underground horse whims were sometimes used in the area, these as at surface requiring a large circular space in which the horse could walk round the central winding drum, the ground having to be cut out for this purpose. Such a whim chamber exists in Smallcleugh mine some 220yd from the portal. The whimsey shaft is located through the arch behind the low wall. The chamber is partially flooded and much water pours from the roof, due, perhaps, to the pervious and jointed rock cover overhead, but generally the inner reaches of the mine are quite dry. (1975)   Portal NY787428

129

129   Towards the inner reaches of Smallcleugh there exists a massive worked-out stope known as the Ballroom Flat. It was here on 2 September 1901 that Mr C. Harper, clerk to the Vieille Montagne Zinc Company and twenty-eight members of a local Masonic Lodge sat down to a meal specially prepared in a local hotel, and trammed into the mine through nearly a mile of haulage levels. Such unique repasts seem to be a feature of metal mining history, for in Cornwall meals were sometimes taken in such curious places as submarine levels and in large pumping engine cylinders prior to their erection. Remains of the straw used on the floor can still be seen in this flat, while broken crockery is occasionally found among the debris. The banquet took place as intimated above during the Vieille Montagne Zinc Company's term of operations, this concern being responsible for continued extension and stoping of the mine for both its zinc and lead content, the mineral being dressed in a newly constructed mill at Nenthead, built in 1910. (1975)   Portal NY787428

130   The Ballroom Flat of Small-cleugh is not the only large underground excavation in the area. The Rotherhope Fell Mine boasts a similar, if less fanciful, chamber or engine room at the point where the principal haulage level reaches the main vein. This engine room was excavated to house hoisting and pumping machinery working in conjunction with two deep underground shafts (winzes) which were sunk to explore the vein in depth. Both these winzes are, of course, flooded to the collar; No 1 is situated near the centre of this chamber behind the pile of debris and is 370ft deep. As in the case of a similar exercise at the Sir Francis Mine, Swaledale, the ore values of the vein in depth were found to be poor. The entire engine room is lined with masonry thus giving the impression of the existence of weak ground at this point, this masonry having the appearance of being painted entirely matt black. A great deal of pipework and a number of valves remain in the area although all trace of the pumping and winding machinery has vanished. (1975)   NY699427

131 Beside the large engine room in Rotherhope Fell Mine and on a level with the main adit is the No 2 shaft, which still has the cages in situ. There was also a third (No 3) shaft sunk to a depth of 140ft, but this was some way to the south west and is now inaccessible. Around this 120ft deep shaft, concrete arching lines the tunnel and also extends into a branch level, this having a hole in the floor suggesting that below exist partly collapsed and flooded stopes. Rotherhope Fell mine was latterly worked, in common with others in the area, by the Vieille Montagne Zinc Company and it is the work of this concern that largely remains accessible and also visible at the level mouth. Power in the dressing mill at this time was derived from a water turbine, while compressed air was supplied from a hydraulic system and electric compressors installed underground. The main adit, the Black Burn Level, was the work of John Smeaton, who held the position of Receiver for the Greenwich Hospital in the 1760s and whose work on planning the great Nent Force Level is well known.
(1975)   Portal NY699427

132 The Brownley Hill or Blooms-bury Mine worked the Brownley Hill, Wellgill, Gudham Gill, and Scaleburn Veins and its levels and workings are very extensive, these stretching from Cumberland into the neighbouring county of Northumberland. It was incorporated into a lease covering some ten square miles around Nenthead and operated by the London Lead Company. Access to the workings was by two levels which connected underground, these being termed the High and Low Levels and commencing in Gudham Gill. The High Level is no longer accessible, but the Low Level (above) still remains open and leads to a maze of ramifying workings of considerable magnitude which have produced over 14,500 tons of lead concentrates together with nearly 2,000 tons of zinc.
(1975)   NY776446

133 This illustration gives some impression of the enormity of the worked-out flats in Brownley Hill Mine. There also remains here a working example of an ore waggon, the design of which varied little throughout most northern lead mines. Other relics include wheelbarrows, shovels and rusted mining implements, as well as a more modern series of cable stanchions and rollers used in connection with a diesel-operated pulley system dating from a small scale twentieth-century reworking of the mine. It may be added that the powerful veins wrought by this mine extended through to West Allendale and there were exploited by the Blackett/Beaumont concern with considerable success.
(1975)   Portal NY776446

132

133

134 In a large stope at the Brown-ley Hill Mine, the rusting body of an ore waggon stands before the entrance of an arched haulageway leading to distant stopes. The constr-uction of such arching deep under-ground is even more amazing when it is considered that the work had to be originally carried out entirely by the feeble illumination of candles. (1975) Portal NY776446

135 Allenheads Mine, at the head of East Allendale, was one of the richest workings in the locality. It belonged to the Blackett/Beaumont concern and was originally worked from surface gin shafts sunk on rising ground above the village. In line with the policy operating at the nearby London Lead Company's mines, the Beaumonts came to realise that horse levels for access were a more practical proposition and Allenheads Mine thereafter was chiefly worked by this latter method. The principal haulage adit was Fawside Level, driven in the 1790s and this takes a general south-westerly course cutting the Old, Wentworth, Henrietta and Great Veins and ends beneath the head of Rookhope Valley near Corbetmea Dam. Its course can be traced on the surface by a line of walled ventilation shafts which stretch up the hillside in railway tunnel fashion. The illustration shows Collier Shaft near the Rookhope-Allenheads road. The mine, now known as Beaumont Mine, is presently producing fluorite for steelmaking purposes. (1975) NY868454

136 and 137 With a view to exploring East Allendale and unwatering Allenheads Mine, the Beaumont Company commenced driving a very ambitious level from below Allendale Town. Construction began in 1855 under the direction of Thomas Sopwith FRS who was General Manager for the Beaumonts. Driving of the level was undertaken from the riverside and concurrently from a series of shafts along the proposed course — the Studdon Dene, Holmes Linn and Sipton Shafts. These shafts opened up some very productive ground and for supplying power for resultant dressing operations and pumping, hydraulic rotary engines were employed. High-pressure water for these was derived from hydraulic accumulators fed from a series of waterwheel-powered pumps driven by the river, this arrangement enabling 'power' to be piped to any location where it was required. In common with many similar schemes, the level never reached its final objective because the Allenheads Mine was closed in 1896, but driving of the tunnel continued until 1912 as an exploratory venture. Its forehead beneath Spartylea stands 4·68 miles from the portal. Named

the Blackett Level in honour of the founder of the Beaumont Company, the entrance is easily reached by footpath beside the river and exhibits the very finest features of horse level design. The masonry arching at its mouth and for hundreds of feet inside where it passes through loose boulder clay, is beautifully executed, profiled as it is, to pass the 'galloways' or level horses working on underground haulage. Even the Blackett Level pales in comparison to similar schemes undertaken overseas, notably the Ernst August Adit in the Harz district of Prussia, which was driven over 14 miles by hand labour and completed in 1864.

However, this English example serves well to illustrate the skill, boldness, and engineering capabilities of generations of metal miners both in the North or elsewhere who, during the past three centuries transformed mining from a hit-and-miss occupation into virtually a science.
(1975)   NY836560

Glossary

Adit: tunnel into a mine

Adventurer: one who invests in mining

Bal maiden: female ore dresser

Blende (ZnS): principal ore of zinc

Blind deposit: concealed ore shoot

Bob: steam engine or balance beam of iron (formerly of timber)

Buddle: ore dressing device

Burrow: Cornish term for waste heap

Cassiterite (SnO$_2$): principal ore of tin

Collar: top of a shaft

Concentrates: dressed ore ready for smelting

Cupola: a smelting furnace on the reverberatory principle

Deads: broken rock containing no ore

Dressing: washing and processing to separate ore from waste

Drift: a tunnel driven through an orebody, or along a vein or lode

Fathom: unit measure of length and depth pertaining to metal mining – six feet

Flats: lateral ore replacement deposits

Flat rods: reciprocating power transmission timbers (or iron rods)

Flue: horizontal or inclined culvert for the conveyance of smoke or fume

Footwall: wall under a vein or lode – the underlying wall

Forehead: end of driven tunnel – the working face (sometimes called forefield)

Galena (PbS): principal ore of lead

Hanging wall: the rock on the upper side of vein or lode

Hutch: storage bay

Kibble: egg-shaped ore hoisting bucket

Launder: timber trough for conveyance of water over unsuitable terrain

Leat: ditch excavated as a water conduit

Lever wall: substantial wall of masonry to carry steam-engine beam

Lode: term for ore shoot (mainly Cornish and Welsh term)

Man engine: mechanised reciprocating rods in mine shaft providing ingress and egress for miners

Pipe vein: large mineral deposit usually filling old caverns

Rake vein: near vertical ore shoot contained within distinct walls – usually faults

Pitwork: pumping rods connected to waterwheel or steam engine

Portal: entrance to tunnel

Rise: vertical shaft rising above working level

Run-in: collapse of an excavation

Sett: area of a mine's lease – its working confines

Sough: drainage tunnel driven to unwater a vein or veins (mainly a Derbyshire term)

Spalling: breaking down veinstuff on the dressing floor to release the mineral

Stamps: mechanised crushing equipment

Stemples: timber (sometimes stone) bars jammed across workings for support and climbing purposes

Stope: the working place in a mine where ore is extracted

Tailings: finely crushed waste material from the dressing process

Tributers: self employed individuals working in groups for a predetermined financial return per unit of ore produced

Underlie: dip or angle of shaft, vein or lode from the vertical (referred to as 'hade' in Derbyshire)

Wheal: Cornish prefix to mine name derived from 'huel' signifying a place of work

Winze: underground shaft – probably derived from 'winds' or downward excavations made to improve ventilation in workings of early date

Bibliography

The following books and articles have been suggested for further consultation by readers wishing to learn more of specific mining fields and individual mines. There are, of course, many other works, no less informative, but to list all such material concerning metal mining in Britain as a whole is too great a task to attempt here. The writer's first publication, *Britain's Old Metal Mines: A Pictorial Survey* (Truro 1974) may be regarded as a companion volume to the present work and contains illustrations and information on areas not mentioned here.

SOUTH-WEST ENGLAND

A comprehensive reference work relating to virtually all mines and trials within Cornwall and Devon is H.G. Dines's *The Metalliferous Mining Region of South West England* (HMSO, 1956) but although thorough this tends to be highly involved and is best read in conjunction with the 1908-9 edition 6in Ordnance Survey maps, copies of which are obtainable from the Maps Department of the British Museum. D.B. Barton's *A Guide to the Mines of West Cornwall* and *The Mines and Mineral Railways of East Cornwall and West Devon* (Truro, 1963 and 1964) are easily assimilated guides to the more important workings in the West Country, while *A History of Tin Mining and Smelting in Cornwall* and *A History of Copper Mining in Cornwall and Devon* (Truro, 1967 and 1968) are further volumes by the same author dealing with the subject in greater depth. On steam pumping engines, which were widely used in most deep mining areas, *The Cornish Beam Engine* by D.B. Barton (Truro 1969) may be regarded as a standard work on the subject, while F.D. Woodall's *Steam Engines and Waterwheels* (Hartington, 1976) is a delightful dissertation on these prime movers and includes many photographs of these machine when they were in use. Pictorial works of a historical nature published in Truro include *Historic Cornish Mining Scenes at Surface* by J.H. Trounson and *Historic Cornish Mining Scenes Underground* edited by D.B. Barton; also two volumes entitled *Cornish Engine Houses: A Pictorial Survey* by H.G. Ordish, portray some of these buildings containing engines prior to their being scrapped. A long overdue biography of Britain's foremost nineteenth-century metal miner and engineer has recently appeared entitled: *John Taylor, mining entrepreneur and engineer 1779-1863* by R. Burt (Hartington 1977) and throws considerable light on this outstanding individual whose skill and success in Cornwall was to spread throughout the British Isles and ultimately to important mining fields overseas. The Trevithick Society, whose prime objective is the study of history and technology in Cornwall, produces occasional volumes related to metal mining in the West and their *Dolcoath:*

Queen of Cornish Mines by T.R. Harris (Camborne 1974) is a first-class example of research into a large and long-lived concern not untypical of South West England. Finally, a book of outstanding interest containing chapters on the life and work of the miner is *The Cornish Miner* A.K. Hamilton Jenkin (reprinted Newton Abbot, 1972).

WALES

The metal mines of South Wales are poorly documented and have received relatively little attention due to their scattered nature and the more numerous and productive mines further North. However, G.W. Hall's *Metal Mines of Southern Wales* (Westbury-on-Severn, 1971) is an excellent reference book on the region. The mines of central Wales are well dealt with in a series of booklets entitled *The Old Metal Mines of Mid Wales* by D.E. Bick (Newent 1974). Material of a more general nature encompassing the whole principality includes *Mining for Metals in Wales* by F.J. North (Cardiff, 1962), *Lead Mining in Wales* by W.J. Lewis (Cardiff, 1967), and *Mines, Mills and Furnaces* by D. Morgan Rees (HMSO, 1969) the latter giving useful information although by no means comprehensive due to space limitations. The Northern Mine Research Society, Skipton, have published two special surveys dealing briefly with every mine in Denbighshire and Flintshire, under the general heading *Non-Ferrous Mines of Wales* by J.R. Foster-Smith and further monographs of this type are planned to appear in the future. Moreover, information on the use of steam pumping-engines in North Wales is also to be found in the society's *Memoirs* for 1968. The more specialized technical *Geological Survey Memoirs* are of unique value to the serious student and contain information on individual mines and geological data. In particular the following will be found invaluable: Vol XX, 'Lead and Zinc, The Mining District of North Cardiganshire and West Montgomeryshire' by O.T. Jones, (HMSO, 1922); *Bulletin of the Geological Survey of Great Britain* No. 14, 'The Mineral Veins of the Minera, Maeshafn District of North Wales' by J.R. Earp (HMSO, 1958).

SHROPSHIRE

Two publications are of interest to the researcher in this locality: *The Shropshire Lead Mines* by F. Brook and M. Allbutt (Hartington, 1973) is a good general introduction to the subject and more technical information will be found in the *Bulletin of the Geological Survey of Great Britain* No 14, 'The West Shropshire Mining Region' by H.G. Dines (HMSO, 1958).

DERBYSHIRE

The lead mining industry here is very well documented and there is much literature for the reader to study. A general outline of the area's mineral field and customs is *Derbyshire Lead Mining Through the Centuries* by N. Kirkham (Truro 1968). A very popular guide is provided by members of the Peak District Mines Historical Society entitled *Lead Mining in the Peak District* (Bakewell, 1968). The latter society also publishes *The Bulletin* half yearly, which incorporates the researches of members on most mines and aspects of the industry. Also by the same society, a re-printed version of A.H. Stoke's work *Lead and Lead Mining in Derbyshire* (1880), contains much historical background material relating to the laws and customs of the area and the state of the industry at the date of original publication. Historic pictorial material is to be found in *Derbyshire's Old Lead Mines and Miners* by J. Rieuwerts (Hartington, 1972) and *The Caverns and Mines of Matlock Bath* by R. Flindall and A. Hayes (Hartington, 1976) is a detailed survey of this small but intensively-mined corner of the county. On a more general scale, A. Raistrick and B. Jenning's *A History of Lead Mining in the Pennines* (1965) gives information on the industry of the Pennine mining field as a whole.

THE ISLE OF MAN

The paucity of printed material available on this area is reflected in the fact that only two major works are worthy of suggested further reading. *Economic Geology of the Isle of Man with Special Reference to Metalliferous Mines* by G.W. Lamplugh was published by the Geological Survey in 1903 and describes all the island's mine workings and trials — a later *Memoir* by the same publisher touches upon the mines but its content is insignificant. More recently, the Manx mines are dealt with in *Industrial Archaeology of the Isle of man* by L.S. Garrad, *et al*, (Newton Abbot, 1972). Casual visitors may find the pamphlet *The Laxey Mines Trail* by F. Cowin (Onchan, 1973) of assistance in interpreting surface features to be found in Glen Mooar and nearby.

YORKSHIRE AND THE NORTH PENNINES

Together with the Raistrick and Jenning's work already mentioned, there is not an inconsiderable quantity of reading matter available appertaining to this region. *Lead Mining in the Mid Pennines* by A. Raistrick (Truro 1973) deals with the mines around Skipton and while the bias is towards early history, it is nonetheless a good introduction to the vicinity. *Mines and T'Miners* by J.M. Dickinson (Sutton-in Craven 1972) is a privately published booklet referring to the history of lead mining in Airedale, Wharfedale and Nidderdale and although a mere 80 pages in length, contains much interesting data together with maps and plans, making it a valuable little reference work on the district. A. Raistrick's *The Lead Industry of Wensleydale and Swaledale*, Vols I & II, (Hartington, 1976) is a comprehensive and well illustrated guide, with the mines dealt with in Vol I and the smelting mills in Vol II. R.T. Clough's *The Lead Smelting Mills of the Yorkshire Dales* published privately (Keighley, 1962) is a book concerned, as the title suggests, with nearly all the smelt mill sites between Wharfedale and Swaledale and includes drawings and photographs, together with information on early mining techniques. Further north the Geological Survey covers the area in *The Northern Pennine Orefield* (HMSO, 1949) and although strictly a reference work, is of use to the student of mining history in that production statistics and data on geological conditions are included. Concerning the life and work of the miner, C.J. Hunt's *The Lead Miners of the Northern Pennines* (Manchester, 1970) gives an insight into the conditions prevailing at the large mines in Nenthead and Allendale, and this volume can be linked with A. Raistrick's *Two Centuries of Industrial Welfare* (Hartington, 1977) the latter being a work dedicated to the activities of the London (Quaker) Lead Company. The Northern Mine Research Society concentrates much of its attention on the Yorkshire and North Pennine mines and their *Memoirs*, together with their *Occasional Publications* are excellent vehicles of information. The latter publications include *The Greenhow Lead Mining Field.* (Occasional Publication No. 4) which is a detailed history of this more southern mining area.

Concerning the country as a whole and containing many interesting items on past activity and future mining prospects is the publication *The Future of Non-Ferrous Mining in Great Britain and Ireland* by the Institute of Mining and Metallurgy (1959). This has been found of great assistance in clearing up points not adequately dealt with in other printed works. Most of the foregoing themselves contain comprehensive bibliographies and reference to these will further aid the reader in locating the immense amount of written material available on the subject.

Index